ACTA NEUROCHIRURGICA
SUPPLEMENTUM 32

Trauma
and Regeneration

Special Symposium
of the 9th International Congress
of Neuropathology, Vienna, September 1982

Edited by
J. Hume Adams

SPRINGER-VERLAG WIEN GMBH

J. HUME ADAMS, M.B., Ch.B., Ph.D., F.R.C.Path., F.R.C.P.

Professor of Neuropathology, University Department of Neuropathology,
Institute of Neurological Sciences, Glasgow, Scotland, U.K.

Secretary-General, International Society of Neuropathology

With 32 Figures

Library of Congress Cataloging in Publication Data. International Congress of
Neuropathology. (9th, 1982, Vienna, Austria) Trauma and regeneration. (Acta
neurochirurgica. Supplementum; 32) 1. Central nervous system—Wounds and
injuries—Congresses. 2. Nervous system—Regeneration—Congresses. I. Adams,
H. (Hume), 1929— . II. Title. III. Series. RD593.I53 1982 617'.48044 83-16950.

ISSN 0065-1419
ISBN 978-3-211-81775-9   ISBN 978-3-7091-4147-2 (eBook)
DOI 10.1007/978-3-7091-4147-2

Preface

The General Council of the International Society of Neuropathology enthusiastically and unanimously endorsed the suggestion made by the Executive Committee—chaired by Professor
Dr. Franz Seitelberger, Vienna—for the IXth International
Congress of Neuropathology that one of the major symposia at
that Congress should be on Trauma and Regeneration of the
Central Nervous System. The reasons for this are not difficult to
understand: non-missile head injury and its sequelae—often a
permanently brain damaged young adult—is one of the major
problems that has faced society for many decades, and is continuing
to do so since relatively little progress appears to have been made in
its prevention; and the hope is that experimentalists may be able to
shed some light at least on the potential for regeneration in the
central nervous system. These proceedings are the outcome of that
very successful symposium held in Vienna in September 1982. The
Society is most grateful to Allgemeine Unfallversicherungsanstalt
(AUVA), Vienna, for their sponsorship.

Of the four major presentations, two were on the subject of non-
missile head injury in man and experimental animals, and two dealt
with recent developments in the field of regeneration. The former
review the clinical features and their structural basis and establish
that all of the major types of brain damage seen in man as a result of
a non-missile head injury have now been reproduced by controlled
angular acceleration of the head in subhuman primates without
anything striking the head. One of the papers on regeneration deals
particularly with scar formation and emphasizes how important the
effects of age are on responses to injury in the brain: the second
shows that interneuronal connections can occur between transplants and adult host brains.

The shorter reports based on platform presentations and posters
cover a very wide range of the types of damage that may occur in the
brain and the spinal cord, and their identification, analysis, and
significance.

These Proceedings contain a wealth of information for anyone
with an interest in any aspect of injury to the central nervous
system.

Glasgow, Scotland, October 1983   J. HUME ADAMS

Contents

Acta Neurochirurgica, Suppl. 32, 1—13 (1983)

Division of Neurosurgery, University of Pennsylvania, Philadelphia, Pennsylvania, U.S.A.

Head Injury in Man and Experimental Animals: Clinical Aspects

By

T. A. Gennarelli

With 1 Figure

Summary

Clinical studies have demonstrated that, with regard to death, the two worst types of head injury are subdural haematoma (SDH) and diffuse axonal injury (DAI). These two have different mechanisms of causation; SDH occurs much more commonly in non-vehicular injuries, especially falls, while DAI is caused, almost exclusively by vehicular mechanisms. The production of these two types of injury in non-impact acceleration models helps to explain these causal differences, but also shows that both injuries share a common mechanical cause, differing only in degree. SDH is due to vascular injury that is caused by relatively short duration angular acceleration loading at high rates of acceleration. These are the circumstances that occur in falls where the head rapidly decelerates because of impact to firm, unyielding surfaces. DAI is also due to angular acceleration of the head, but occurs most readily when the head moves coronally and it only occurs when the acceleration duration is longer and the rate of acceleration lower than conditions that produce SDH. These conditions are met in vehicle occupants where impact to deformable or padded surfaces lengthens the deceleration and decreases its rate. In DAI the principal mechanical damage is to the brain itself (mainly to axons) while in SDH the primary damage occurs to surface blood vessels.

Now that models of the two most important types of head injury have been created in the laboratory, it is hoped that a better understanding of their pathophysiology will result in new strategies to affect protection from their occurrence and in improved treatment when they do occur.

Keywords: Head injury; man; experimental; diffuse.

Introduction

In general terms, head injuries include two distinct varieties, each having its own unique set of mechanical aetiology, clinical symptomatology, pathophysiology and outcome. *Focal injuries* result from primary localized tissue damage. The cortical contusions, subdural haematomas, epidural haematomas and intracerebral haematomas that comprise the focal injuries can cause coma if they are sufficiently large to produce brain shift, herniation and brain stem compression. The *diffuse brain injuries* are fundamentally different and are associated with widespread primary brain damage. This damage may be principally functional as in the case of concussive injuries, or may be structural as seen in prolonged traumatic coma unassociated with mass lesions, a condition recently termed diffuse axonal injury (DAI)[2,6]. The diffuse brain injuries, when sufficiently severe, cause coma, not by brain stem compression, but by primary injury to the cerebral hemispheres or the brain stem.

Our recent efforts have been aimed at 1) defining the importance of the different types of head injury with regard to death and disability, 2) determining the circumstances that cause the injuries and 3) duplicating these injuries in experimental models. This paper reviews these efforts.

Importance of Lesion Type to Mortality and Morbidity

Data from University of Pennsylvania Head Injury Center

In order to ascertain the relative importance of the various types of head injury, 434 hospitalized head-injury patients were analyzed. These patients comprised consecutive admissions to the University of Pennsylvania Head Injury Center and all had at least one computerized tomographic (CT) scan. Patients of all degrees of severity were included and each diagnosis (lesion) obtained from CT scan, surgery, necropsy or clinical sources was recorded. Table 1 records the most frequently occurring lesions that were judged most responsible for the clinical symptoms and signs, their incidence in the series, their outcome and their incidence in those who died. Acute subdural haematoma (SDH) occurred in 9% of the 434 patients, was fatal in 47% of these patients, and accounted for 34% of the deaths. The severe form of DAI (coma > 24 hours with brain-stem signs) was similarly over-represented in the non-survivors. Although present in only 4% of the whole group, its high

Table 1. *Importance of Lesions with Regard to Outcome
all Hospitalized Patients.* n = 434

Lesion type	Overall incidence %	Outcome good-mod. %	Outcome severe-veg. %	Outcome death %	Incidence in non-survivors %	Mortality index = deaths per 10,000
SDH	9	34	19	47	34	423
Severe DAI	4	6	29	65	21	260
Moderate DAI	13	33	56	11	11	143
ICH	4	50	22	28	9	112
Contusion	13	82	11	7	7	91
EDH	5	68	28	4	2	20
Concussion	24	97	3	0	0	0
Miscellaneous	28	79	14	7	16	196
Total	100				100	1,245

For this and subsequent tables and figures the following abbreviations are used:
SDH = acute subdural haematoma.
DAI = diffuse axonal injury (prolonged coma without mass lesions).
   Severe DAI = coma > 24 hours with decerebration and brain-stem signs.
   Moderate DAI = coma > 24 hours without decerebration.
ICH = intracerebral haematoma.
EDH = extradural haematoma.
Outcome categories are according to the Glasgow outcome scale.
Mortality index = (overall incidence) × (mortality rate).
fx = fracture.
Outcomes are % of each lesion type (for each lesion % good-mod. + % severe-veg. + % dead = 100).

mortality rate (65%) resulted in 21% of the deaths. Next in frequency as a contributor to death was the moderate form of DAI (coma > 24 hours without brain-stem signs). This lesion occurred in 13% of the patients, caused death in 11% and accounted for 11% of the deaths in the series. Thus two lesions, SDH and DAI, were responsible for two thirds of the deaths, more than all other lesions combined.

The least severe injuries were those associated with a high incidence of acceptable survival (good and moderate recovery). Table 1 shows that the most favorable head injuries are concussion, contusion and extradural haematoma with 97, 82, and 68% attaining acceptable recovery respectively.

In determining the overall significance of a single diagnosis within the entire spectrum of lesions, consideration must be given not only to the mortality rate of the lesion, but also to how frequently that lesion occurs. Thus, a lesion may cause a high percentage of fatalities, but is so uncommon that the total number of deaths that result is small. On the other hand, the mortality rate for a lesion may be moderate, but the lesion is so frequent that many deaths are caused by it. Thus, the importance of any particular lesion can be estimated by the product of its frequency (incidence) and its mortality rate. This product, the *mortality index,* can then be used to determine the overall importance of each lesion in a population of head-injured patients. The mortality index establishes the lesion's importance for a series of 10,000 similarly injured patients and expresses the number of patients who will die of a particular lesion in a series of 10,000. Thus for SDH, severe DAI and moderate DAI respectively, Table 1 predicts that for 10,000 hospitalized head-injured patients, 423 will die with SDH, 260 from severe DAI and 143 from moderate DAI. Conversely, traumatic intracerebral haematoma (ICH) had the same incidence as severe DAI (4%), but because its mortality rate was lower (28%), ICH is not as important a lesion as is DAI (mortality indices of 112 and 260 respectively).

These data demonstrate, overwhelmingly that *the two most important lesions with respect to death from head injury are SDH and DAI.*

Data from Multicenter Study of Outcome

Data were gathered from seven head-injury centres in the United States to test the results found in the University of Pennsylvania series[7]. Because the intention was to determine the conditions most responsible for death and disability, this study was confined to severely injured patients. Thus this multicenter study comprised all patients with non-missile injuries who had Glasgow coma scores (GCS) less than nine for at least six hours whereas the Pennsylvania series referred to above included all hospitalized head-injured patients. Table 2 summarizes the findings. There were 1,107 patients in the series, 56% with focal injuries and 44% with diffuse brain injuries. Not only were focal injuries more frequent than diffuse injuries, but overall, focal injuries were more lethal than were the diffuse injuries, 48% and 32% mortality respectively. As in

Table 2. *Importance of Lesions to Mortality in Severe Head Injury*

Lesion type	N	Incidence (%)	Lesion mortality (%)	Mortality index
Focal injuries				
Extradural haematoma	96	9	20	180
Subdural haematoma	319	29	61	1,769
Other focal injuries	205	18	39	702
Diffuse injuries				
DAI-mild (coma 6–24 hours)	92	8	15	120
DAI-moderate (coma < 24 hours; not decerebrate)	219	20	24	480
DAI-severe (coma < 24 hours; decerebrate)	176	16	51	816
Total	1,107	100	41	4,067

the Pennsylvania series, SDH was the worst lesion in head injury; it carried a 61% mortality (74% in those with a GCS 3, 4, or 5) and because of its high incidence (29% of the series), it alone was responsible for 43.5% of all deaths. Moderate and severe DAI together were more common (36% of the patients), but the mortality rate was less (36%). They were however responsible for 32% of the deaths. Thus two lesions, DAI and SDH accounted for three fourths of all deaths.

This multicenter series of severely head-injured patients substantiated the findings of the University of Pennsylvania series, *i.e. SDH and DAI produce more head-injury deaths than all other lesions combined.*

Causal Circumstances Associated with Various Head Injuries

The circumstances that caused the various types of lesion were analyzed from the University of Pennsylvania series. Table 3

Table 3. *Causes of Severe Head Injuries (GCS 8 or less)*

Injury cause	Focal injury	Diffuse injury	
Vehicle occupant	19%	41%	
Pedestrian	14%	40%	
Fall	48%	10%	
Assault	16%	8%	
Other	3%	1%	
	100%	100%	

Injury cause	Focal injury	Diffuse injury	Total
Vehicle occupant	24%	76%	100%
Pedestrian	20%	80%	100%
Assault	56%	44%	100%
Fall	73%	27%	100%

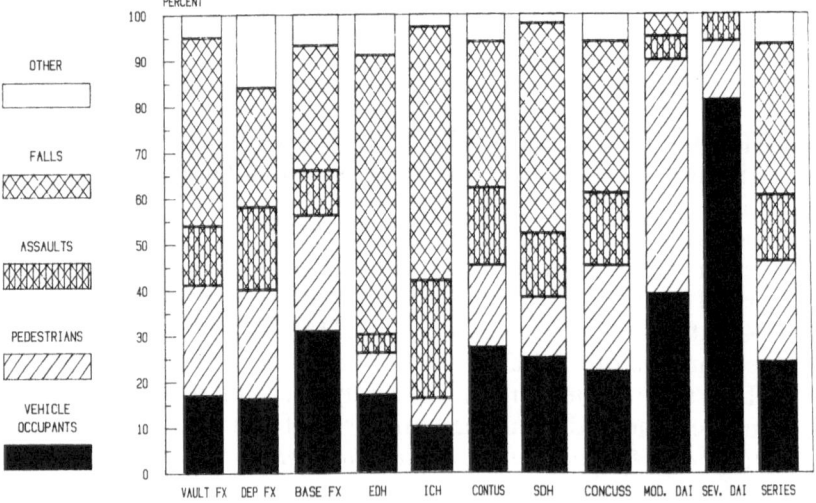

Fig. 1. Causes of common types of head injury (all hospitalized patients, n = 434)

Table 4. *Frequency of Common Injuries According to Cause*

	Less common than expected	As common as expected	More common than expected
Vehicle occupants	* ICH vault fx depressed fx EDH	concussion contusion SDH	* severe DAI moderate DAI basilar fx
Pedestrians	* ICH EDH SDH severe DAI contusion	concussion vault fx depressed fx	* moderate DAI basilar fx
Assaults	* EDH basilar fx moderate DAI severe DAI	concussion contusion vault fx SDH	* ICH depressed fx
Falls	* severe DAI * moderate DAI depressed fx basilar fx	concussion contusion	* EDH vault fx SDH ICH
Other	* moderate DAI * severe DAI SDH ICH	concussion contusion vault fx basilar fx	* EDH depressed fx

* $p = > 0.05$.

demonstrates the overall findings in the severely injured patients (GCS < 9). There is a marked difference in the type of injury produced by various causes. Whereas almost half (48%) of the focal injuries were caused by falls, only 10% of the diffuse injuries were produced by this mechanism. Conversely, vehicular injuries (occupants and pedestrians) accounted for 81% of the diffuse injuries and for only one third of the focal injuries.

Table 3 also shows that vehicular accidents produced diffuse injuries three to four times as often as focal injuries while the inverse relation exists for falls (1:3, diffuse:focal). Assault injuries caused an almost equal number of focal and diffuse injuries.

The cause of injury for the more frequent specific injury types in the whole series is shown in Fig. 1. If each cause of injury produced the same spectrum of injuries, it would be expected that for each specific injury the distribution of causes would be exactly the same as for the whole series (the furthest right column in Fig. 1). Thus, since vehicle occupant mechanisms caused 24% of the injuries to the 434 patients, it is expected that 24% of each of the specific injuries would be due to this cause if occupant injuries were similar to the other injury mechanisms. That this is not the case is demonstrated in Fig. 1, where the frequency of occupants is much higher than expected for severe DAI, moderate DAI and basilar fracture. Similarly pedestrians are much more likely to incur moderate DAI than expected. Only concussion and cortical contusion are caused by mechanisms with frequencies equal to that anticipated by the distribution of causes. Table 4 extracts the frequency of the injuries according to the causes of injury and shows that vehicular mechanisms (occupant and pedestrian) are much more likely to cause DAI whereas assaults, falls and other mechanisms cause fewer cases of DAI than expected. SDH is more commonly due to falls and is produced less frequently in pedestrian injuries.

In the severely injured patients, *analysis of the two worst head injuries (SDH and DAI) showed a marked difference in cause:*

	SDH	DAI
Falls/assaults	71%	10%
Vehicular	24%	89%

Thus almost all cases of DAI were caused by vehicular injury while most subdural haematomas were non-vehicular in nature.

Experimental Models of Brain Injury

Our goal has been to create the important types of head injury in the laboratory so that these lesions can be more fully studied than is possible in clinical situations. Three groups of experiments are briefly described that have led us to our current state. All three experiments had similar designs and were intended to isolate various injury mechanisms in as pure a form as possible in the laboratory. In order to assure reproducibility from animal to animal certain conditions had to be met: 1) the input to the head must be applied in a quantitatively repeatable, precise manner, and 2) the response of the head must be exactly the same from experiment to experiment. In order to accomplish these ends, sophisticated head accelerating machines were designed that were capable of delivering a programmable and reproducible acceleration load to the head in a manner that prevented impact to the head. The head was coupled to the accelerating machine so that it moved exactly the same as programmed by the machine. Finally, the movement of the head was controlled so that the type and magnitude of head motions were the same from animal to animal.

HAD—II Experiments

The first experiments utilized the head accelerating device-model HAD-II[11]. The HAD-II was fitted to deliver equivalent accelerations of one of two types. Translational acceleration moved the head from posterior to anterior along a straight line, keeping within the anatomical constraints of the squirrel monkey. The second movement was angular acceleration with a centre of rotation in the lower cervical spine. The acceleration in both cases was evenly distributed over the head so that focal loading did not occur (impact was prevented) and the physiological responses to various levels of acceleration were observed. These experiments demonstrated that *angular acceleration regularly produced cerebral concussion whereas translational acceleration, even at higher acceleration levels, did not.* Translational acceleration did produce pathological changes, mostly contusional and occasionally an intracerebral haematoma, but did not produce concussion.

Penn I Device

Our aim was to produce prolonged coma with DAI and because we felt that this condition was a more severe form of injury than concussion we felt it was necessary to increase the acceleration levels from those used in the first HAD-II device animals. However, we also wished to use animals with a larger brain so that more intensive physiological monitoring could be done. This was not physically possible with the HAD-II device so the Penn-I machine was developed. This device used a pneumatic actuator to push a lifter, to which a macaque's head was secured in a plastic helmet. Since our interest was in the concussive injuries and since the HAD-II studies showed that translational acceleration did not produce these injuries, only angular acceleration was used.

By moving the head 60° from posterior to anterior in the sagittal plane, varying degrees of injury were produced depending on the amount of acceleration used[1, 8, 9]. An experimental trauma severity (ETS) scale was devised to describe these degrees of injury and was based on observable changes in cardiac, respiratory or neurological status following injury. Seven responses to acceleration were found as follows:

In ETS grade 0, there was no evidence that the brain had been disturbed by the acceleration. The animals were completely normal behaviourally and neurologically immediately after the acceleration.

In ETS grade 1, there were changes in the systemic arterial pressure or heart rate, but the animals were behaviourally and neurologically normal.

In grade 2, a brief period of respiratory irregularity or apnoea accompanied the alterations of arterial pressure and heart rate, but the animals were not behaviourally unconscious.

Animals with ETS grade 3 had an absent corneal reflex for 30 seconds or less and, in addition, they showed the same changes in vital signs as in ETS grade 2. However, these animals were unconscious for a brief period of time. On awakening, almost always within one to three minutes after the acceleration, behaviour rapidly returned to normal and remained so.

Grade 4 ETS was distinguished from grade 3 by abolition of the corneal reflex for more than 30 seconds. The animals were behaviourally unconscious for 5 to 15 minutes and then awakened slowly. Behavioural abnormality often persisted, manifested as timidity and disinterest in the environment.

Grade 5 was defined as a fatal outcome from a neurological death within six hours of injury. Many animals, however, would have died without life support within a few moments of injury. These animals were instantaneously unconscious and remained so.

Grade 6 ETS was reserved for two animals that received very high acceleration levels. Both suffered instantaneous death and cessation of all vital functions at the moment of injury.

The Penn-I model, using a single injury mechanism, angular acceleration, was therefore able to produce a spectrum of brain injury ranging from no injury to instantaneously lethal. Pathologically—as will be shown in the next paper—almost every type of primary brain injury seen in man was seen in this model [3–5]. The principal exception was DAI, the lesion we most wanted to produce. However, in addition to producing subconcussive injuries (ETS grades 0–2) and several degrees of concussion (ETS grades 3 and 4), *the Penn-I device did produce the most important head injury lesion, acute SDH.* Small amounts of subdural blood were seen in many ETS grade 4 animals, but virtually all ETS grade 5 animals died because of large subdural haematomas.

Penn-II Device

The cause of the subdural haematomas created by the Penn-I device was postulated to be due to the rapidly applied acceleration characteristic of the helmet system and acceleration pulse used in those experiments (10). The Penn-II accelerator was thus created by

Table 5. *Presence of Diffuse Axonal Injury and Duration of Coma, Outcome, and Direction of Head Motion in Subhuman Primates Subjected to the Penn-II Device*

	DAI absent	DAI present
A. Severity of coma		
Cerebral concussion	100%	0
Prolonged traumatic coma		
Mild (16–119 minutes)	50%	50%
Moderate (2–6 hours)	0	100%
Severe and PC (> 6 hours)	0	100%
B. Outcome		
Good recovery	100%	0
Moderate disability	43%	57%
Severe diability and PC	0	100%
C. Direction of acceleration		
Sagittal	92%	8%
Oblique	50%	50%
Lateral	12%	88%

DAI: diffuse axonal injury; PC: prolonged coma at sacrifice.

modifying Penn-I so that the acceleration could be applied more slowly and the head could be controlled more precisely in a metal helmet. Thus the duration of acceleration was longer in Penn-II and the acceleration *rate* was lower. Angular acceleration was used and impact to the head prevented as before.

Lengthening the acceleration duration and decreasing the acceleration rate resulted in traumatic coma that was longer than in Penn-I and, as importantly, did not produce SDH. Finally, the direction of head movement was modified to compare sagittal, 30° off-sagittal oblique and coronal (lateral) motions. It was found that as the direction of head motion went from sagittal to oblique to coronal, there was an increasing duration and severity of traumatic coma. The full lateral motions produced deep coma with decerebration that lasted as long as one week, accompanied—as will be shown in the next paper—by the characteristic pathological findings of DAI[6]. The presence and amount of axonal damage paralleled the severity of injury as measured by the duration of coma, neurological signs and outcome (Table 5). These findings led to the conclusion that axonal damage produced by coronal plane angular acceleration is a major cause of prolonged traumatic coma and its sequelae and that, in the absence of compounding secondary complications, the outcome from head injury depends on the amount and distribution of axonal damage.

It now appears that, except for skull fracture and epidural haematoma, *virtually all types of primary head injury can be produced by angular acceleration applied to the head in the appropriate manner.*

Acknowledgement

This work has been supported by grants from the National Institute of Neurological and Communicative Disorders and Stroke (NS 08803-103) and by contracts from the National Highway Traffic Safety Administration (DOT-HS-9-02088 and DOT-DTNH-22-82-C-07186).

References

1. Abel, J., Gennarelli, T. A., Segawa, H., Incidence and severity of cerebral concussion in the rhesus monkey following sagittal plane angular acceleration. In: Proceedings of 22nd Stapp Car Crash Conference. New York, Society of Automotive Engineers, 1978, pp. 33—53.
2. Adams, J. H., Graham, D. I., Murray, L. S., Scott, G., Diffuse axonal injury due to nonmissile head injury in humans: An analysis of 45 cases. Ann. Neurol. *12* (1982), 557—563.
3. Adams, J. H., Graham, D. I., Gennarelli, T. A., Neuropathology of acceleration-induced head injury in the subhuman primate. In: Seminars in neurological surgery. Proceedings of 4th Annual Conference on Neural Trauma (Grossman, R. G., ed.), pp. 141—152. New York: Raven Press. 1982.
4. Adams, J. H., Graham, D. I., Gennarelli, T. A., Acceleration induced head injury in the monkey: II. Neuropathology. Acta Neuropathol. (Berl.), Suppl. 7 (1981), 26—28.
5. Adams, J. H., Gennarelli, T. A., Graham, D. I., Brain damage in non-missile head injury: Observations in man and in subhuman primates. In: Recent advances in neuropathology (Smith, R., Cavanagh, J., eds.), pp. 165—190. Edinburgh: Churchill. 1982.

6. Gennarelli, T. A., *et al.*, Diffuse axonal injury and traumatic coma in the primate. Annals of Neurology *12* (1982), 564—574.

7. Gennarelli, T. A., Spielman, G. M., Langfitt, T. W., *et al.*, Influence of the type of intracranial lesion on outcome from severe head injury: A multicenter study using a new classification system. J. Neurosurg. *56* (1982), 26—32.

8. Gennarelli, T. A., Segawa, H., Wald, U., Marsh, K., Thompson, C., Physiological response to angular acceleration of the head. In: Seminars in neurological surgery (Grossman, R., ed.), pp. 129—140. New York: Raven Press. 1982.

9. Gennarelli, T. A., Adams, J. H., Graham, D. I., Acceleration induced head injury in the monkey: The model, its mechanical and physiological correlates. Acta Neuropathol. (Berl.), Suppl. *7* (1981), 23—25.

10. Gennarelli, T. A., Thibault, L. E., Biomechanics of acute subdural hematoma. J. Trauma *22* (1982), 680—686.

11. Ommaya, A. K., Gennarelli, T. A., Cerebral concussion and traumatic unconsciousness. Correlation of experimental and clinical observations on blunt head injuries. Brain *97* (1974), 633—654.

Author's address: T. A. Gennarelli, M.D., Associate Professor of Neurosurgery, Division of Neurosurgery, University of Pennsylvania, 3400 Spruce Street, Philadelphia, PA 19104, U.S.A.

Acta Neurochirurgica, Suppl. 32, 15—30 (1983)

* Department of Neuropathology, Institute of Neurological Sciences, Glasgow, Scotland, and ** Division of Neurosurgery, University of Pennsylvania, Philadelphia, Pennsylvania, U.S.A.

Head Injury in Man and Experimental Animals: Neuropathology

By

J. H. Adams*, D. I. Graham*, and T. A. Gennarelli**

With 9 Figures

Summary

All of the principal types of brain damage that occur in man as a result of a non-missile head injury, viz. cerebral contusions, intracranial haematoma, raised intracranial pressure, diffuse axonal injury, diffuse hypoxic damage, and diffuse swelling have been produced in subhuman primates subjected to inertial, *i.e.* non-impact, controlled angular acceleration of the head through 60° in the sagittal, oblique and lateral planes.

Keywords: Head injury; man; experimental; diffuse axonal injury.

Introduction

The most important factor governing outcome in a non-missile head injury—and indeed in almost any type of injury—is the damage sustained by the brain[2]. Some of this may occur at the moment of injury (primary brain damage) but often much of the brain damage is caused by a complication of the original injury (secondary brain damage). Thus, the initial injury—whether mild or severe—may set in motion a progressive and dynamic sequence

Table 1. *Incidence of Focal and Diffuse Types of Brain Damage After Non-Missile Head Injury in Man and Experimental Subhuman Primates*

	Man without DAI (n = 132)	Penn. 1 (n = 53)	Man with DAI (n = 45)	Penn. 2 (n = 26)
Focal brain damage				
Total mean contusion index	17.8	4.2	8.3	1.3
Subdural haematoma	88.0 (66.7%)	19.0 (35.9%)	5.0 (11.1%)	Nil
Raised intra-cranial pressure	114.0 (86.4%)	31.0 (58.5%)	25.0 (55.6%)	Nil
Diffuse brain damage				
Diffuse axonal injury	Nil	Nil	45.0 (100%)	18.0 (69.2%)
Hypoxic damage in cerebral cortex	61.0 (47.2)	4.0 (7.5%)	22.0 (48.9%)	Nil
Brain swelling	24.0 (18.2%)	13.0 (24.5%)	7.0 (15.6%)	Nil

of events, the identification and understanding of which is the essence of the management of a patient with a head injury. There is, however, an increasing tendency to think of brain damage in head injury as being *focal* or *diffuse*[3]. In this era of computerized tomography focal brain damage—and its type—are usually known to be present during life, and are easy to identify post mortem: focal brain damage includes contusions, intracranial haematoma, shift and herniation of the brain, and raised intracranial pressure. In an unconscious patient without any evidence of intracranial haematoma—a situation that occurs in almost 50% of patients who have sustained a severe non-missile head injury[11]—it is usually concluded that the patient has sustained diffuse brain damage but its precise type is rarely identifiable during life: the three principal types of diffuse brain damage that occur in patients who survive their injury for more than a few hours are diffuse axonal injury, diffuse hypoxic brain damage and diffuse brain swelling[3].

There has in the past been a tendency to assume that something must strike the head, or that the head must strike something, before structural brain damage can be brought about by a non-missile head injury. The aim of this brief communication is to demonstrate that all of the types of brain damage that occur in man as a result of a non-missile head injury can be reproduced in subhuman primates without anything striking the head using the Penn. 1 and Penn. 2 devices described in the previous paper[10] and based on inertial, *i.e.* non-impact, controlled angular acceleration of the head through 60° in the sagittal, oblique or lateral planes. As has been shown by Gennarelli in the previous paper[10], the two most important causes of death from non-missile head injury in man are subdural haematoma (SDH) and diffuse axonal injury (DAI), and both of these have now been reproduced in subhuman primates in the laboratory by the Penn. 1 and Penn. 2 devices respectively. In the description that follows a comparison will be made of the principal neuropathological findings (Table 1) in patients dying as a result of a non-missile head injury with and without DAI[6] and in subhuman primates subjected to angular acceleration in the Penn. 1[5] and Penn. 2[12] devices.

Focal Brain Damage

Cerebral Contusions

Contusions occur characteristically at the frontal and temporal poles and on the under aspects of the frontal and temporal lobes where brain tissue comes in contact with bony protuberances in the base of the skull. They are most severe at the crests of gyri but may extend through the cortex into the subcortical white matter: in the early stages they are haemorrhagic and swollen but with the passage of time come to be represented by shrunken, brown scars. Since the damage is focal, patients with quite severe contusions may make a smooth and uneventful recovery from head injury if they do not sustain any other type of focal or diffuse brain damage.

Contusions have long been recognized as a hallmark of head injury, but in previous years assessment of their severity has been made on a descriptive and subjective basis. In an attempt to assess contusional damage quantitatively and more objectively, we have established a numerical system, the *contusion index,* for grading the depth and extent of contusions in various anatomical locators of the brain[7,9,16]. The higher the total contusion index (TCI) in a

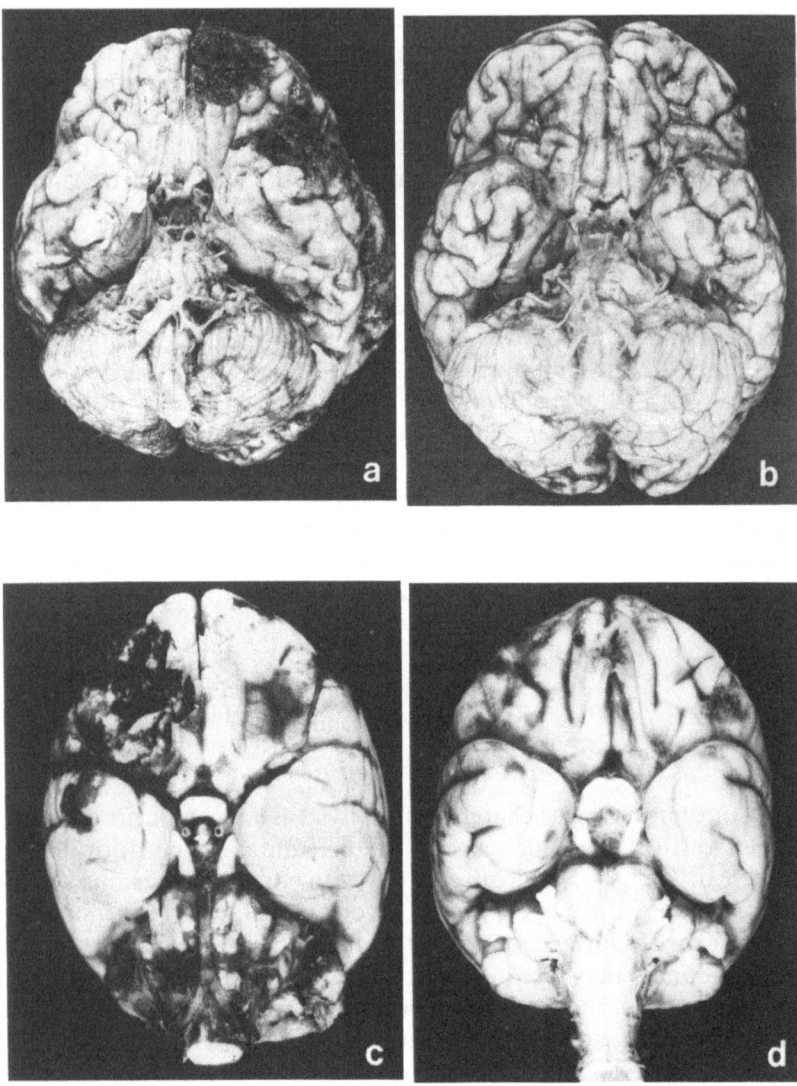

Fig. 1. Cerebral contusions. *a* Patient with subdural haematoma; *b* patient with diffuse axonal injury; *c* subhuman primate subjected to the Penn. 1 device; *d* subhuman primate subjected to the Penn. 2 device. There are haemorrhagic contusions affecting the orbital gyri and the temporal poles. Note that the contusions are more severe in *a* and *c* than in *b* and *d* (*c* from Adams *et al.*[4])

given brain, the greater the severity of contusional damage. This has allowed us to establish (Table 1) that contusions are less severe in patients with DAI (mean TCI = 8.3) than in those without DAI (mean TCI = 17.8; p > 0.0001[6]). Contusions were readily produced in the SDH-inducing Penn. 1[4] device (mean TCI = 4.2) but they were less severe in the DAI-inducing Penn. 2 device (mean TCI = 1.3) (Figs. 1 a–d).

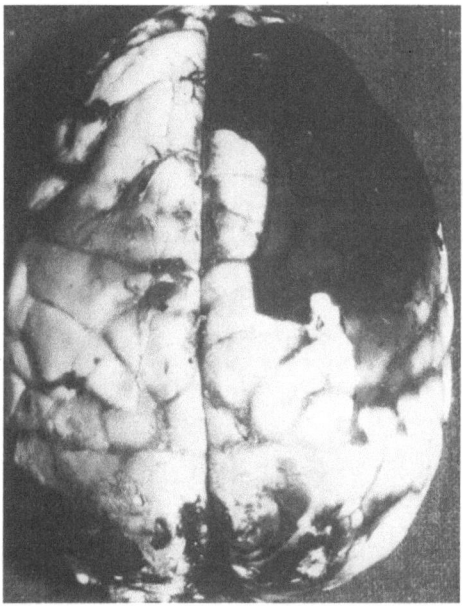

Fig. 2. Acute subdural haematoma covering the anterior two thirds of the right cerebral hemisphere in a subhuman primate subjected to the Penn. 1 device

Intracranial Haematoma

These occur in three principal sites—the extradural space, the subdural space, and within the brain[7]. Extradural haematoma was not produced in subhuman primates with either the Penn. 1 or Penn. 2 devices but this is not surprising since extradural haematoma is basically a complication of a fracture of the skull which, in turn, tears a meningeal artery: fractures of this type were not produced in the experimental model.

Acute SDH often attributable to tearing of parasagittal bridging veins was more commonly present in patients without DAI than in patients with DAI (66.7% compared with 11.1%, p > 0.001; see

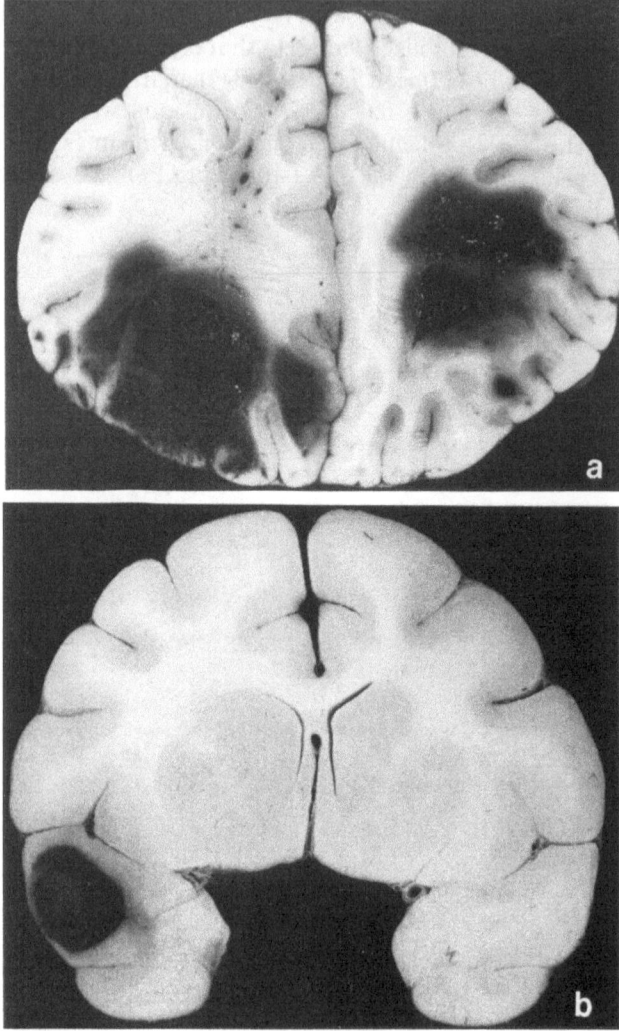

Fig. 3. Intracerebral haematoma. *a* Bilateral frontal haematomas in man without diffuse axonal injury; *b* haematoma in the left temporal lobe of a subhuman primate subjected to the Penn. 1 device (*b* from Adams *et al.*[5])

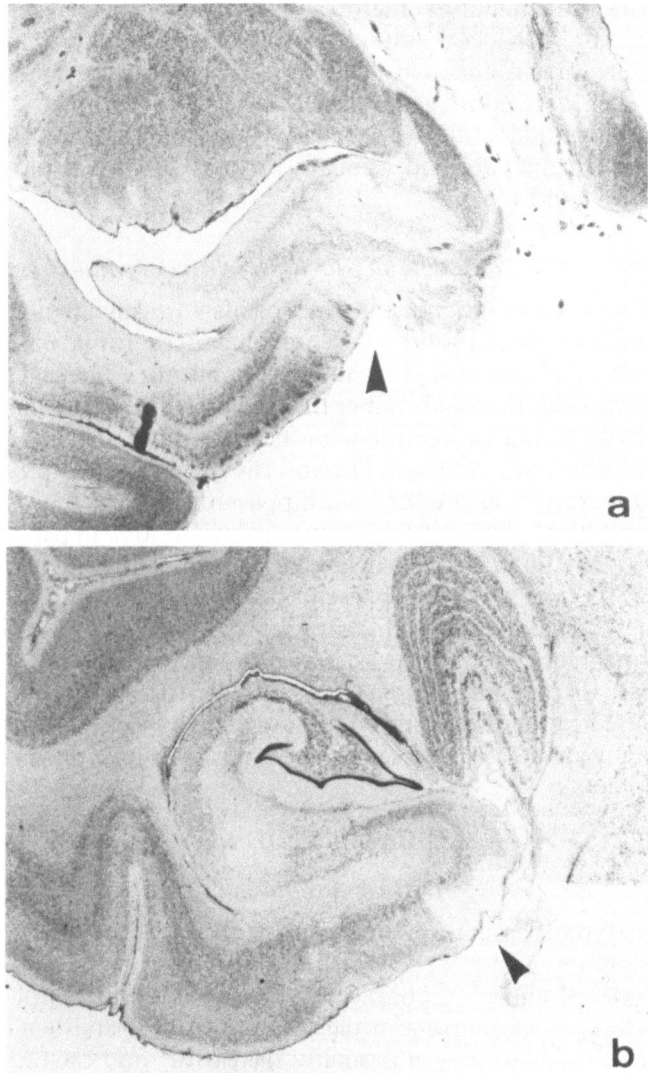

Fig. 4. Pressure necrosis in the left parahippocampal gyrus (arrow). *a* Patient with SDH; *b* subhuman primate subjected to the Penn. 1 device. Note also the subtotal loss of neurons in the Sommer sector of the Ammon's horn. Cresyl violet × 3

Table 1). Intracranial haematoma—usually acute SDH (Fig. 2) was also readily produced with the Penn. 1 device (35.9%): none, however, was produced by the Penn. 2 device. The animals with SDH died very soon after their injury as a result of a rapidly expanding intracranial mass. Intracerebral haematomas similar to those seen in man were occasionally produced by the Penn. 1 device (Figs. 3 a and b).

Brain Damage Secondary to Raised Intracranial Pressure

This is a frequent occurrence in patients who sustain non-missile head injuries, the sequence of events being intracranial haematoma or brain swelling, brain shift, the development of internal herniae, compression of the midbrain or the medulla oblongata, and finally haemorrhage and/or infarction in the brain stem. The incidence of raised intracranial pressure (Table 1) as shown by the presence of pressure necrosis in one or both hippocampal gyri (Fig. 4 a [1]) was 86% in patients without DAI compared with 56% in patients with DAI ($p > 0.001$). Some degree of brain shift and tentorial herniation (Fig. 4 b) was apparent in a few of the Penn. 1 animals which were maintained on life support systems for some hours after the appearance of an acute subdural haematoma, but secondary damage to the brain stem was not identified. This could be attributed to the short period of survival. In none of the Penn. 2 animals was there evidence of raised intracranial pressure.

Diffuse Brain Damage

Diffuse Axonal Injury

This type of brain damage in man has been known in the past by a variety of names [8,15,17-19] but it is a well-defined clinico-pathological entity [6,8] characterized in its classical form by the immediate onset of coma at the time of injury; persistent coma, a vegetative state or severe disability thereafter; and the presence of focal lesions in the corpus callosum (Fig. 5 a) and in the dorsolateral quadrant or quadrants of the rostral brain stem (Fig. 6 a) which can more often than not be identified macroscopically, and histological evidence of diffuse injury to axons. The nature of the latter depends on the period of survival: when survival is short (days) there are numerous axonal retraction balls throughout the white matter (Fig. 7 a): when it is longer (weeks) the most striking feature is the presence of vast numbers of small clusters of microglia throughout

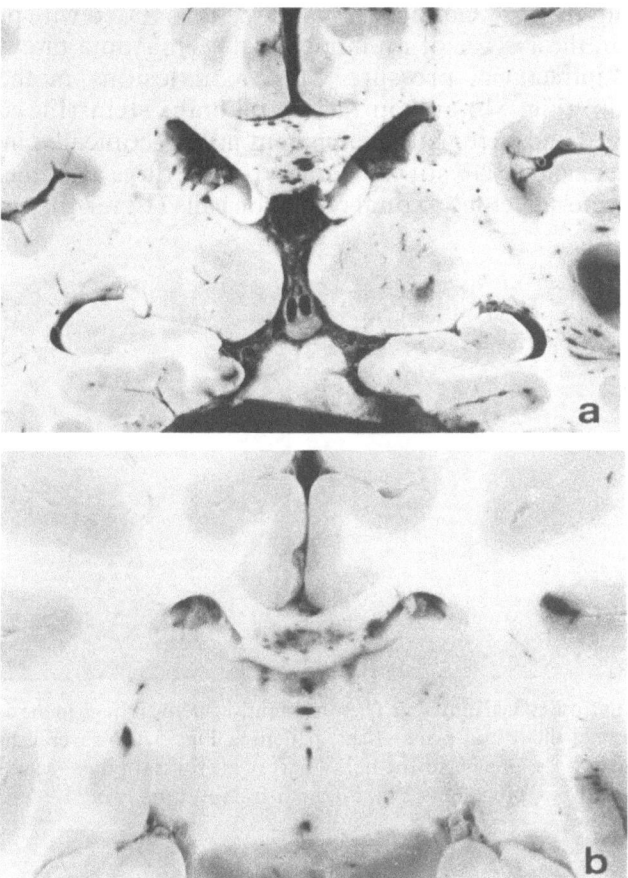

Fig. 5. Diffuse axonal injury. *a* Numerous small foci of haemorrhage are present in the corpus callosum in a patient who died 5 days after a road traffic accident; *b* haemorrhagic lesions in the corpus callosum of a subhuman primate subjected to the Penn. 2 device and sacrificed 48 hours later (*a* from Adams *et al.*[6], and *b* from Gennarelli *et al.*[12])

the white matter (Fig. 8 a); and when a patient survives vegetative for some months after injury[7, 8, 14, 17], widespread destruction of myelinated fibres can be demonstrated by the Marchi technique in the cerebral hemispheres and in descending tracts throughout the brain stem and the spinal cord.

This type of brain damage has been produced for the first time in an experimental model using the Penn. 2 device referred to in the previous paper by Gennarelli[10] and was associated with prolonged coma in the absence of an intracranial haematoma or evidence of raised intracranial pressure. Thus focal lesions in the corpus callosum (Fig. 5 b) and in the rostral brain stem (Fig. 6 b) were produced and were usually apparent macroscopically: in animals that were allowed to survive for up to a few days after their injury, there were numerous axonal retraction balls (Fig. 7 b), particularly

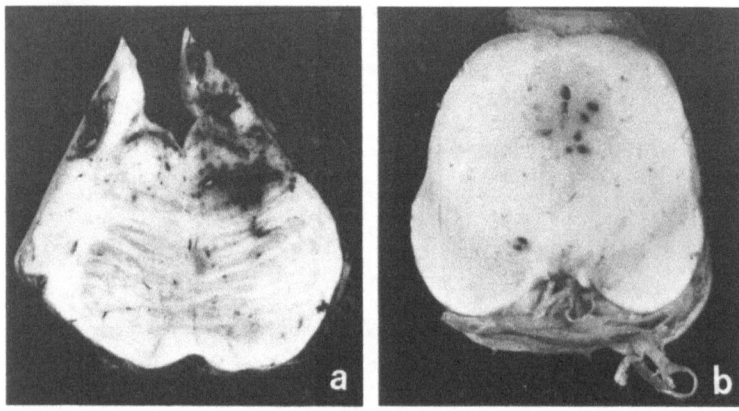

Fig. 6. Diffuse axonal injury. *a* There is extensive haemorrhage in the dorsolateral quadrant of the rostral pons—same patient as Fig. 5 *a*; *b* several small foci of haemorrhage are present in the dorsal part of the rostral pons—same subhuman primate as Fig. 5 *b* (both from Adams *et al.*[6])

in the parasagittal white matter, in the corpus callosum, and in the brain stem; in the animals that were kept alive for more than one week, small clusters of microglia were readily identifiable in white matter (Fig. 8 b): no animal has yet been kept alive long enough to allow of the demonstration of Wallerian-type degeneration of white matter. There was a close correlation between the severity of diffuse axonal injury and the duration of traumatic coma[12].

Hypoxic Brain Damage

This is found frequently in patients dying as a result of a non-missile head injury provided appropriate histological studies are undertaken[13]. Although damage is most frequently seen in the basal ganglia and in the hippocampus, there is also a high incidence

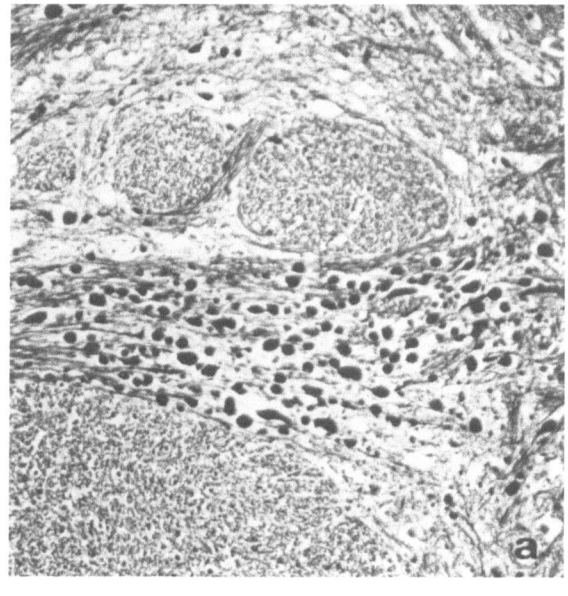

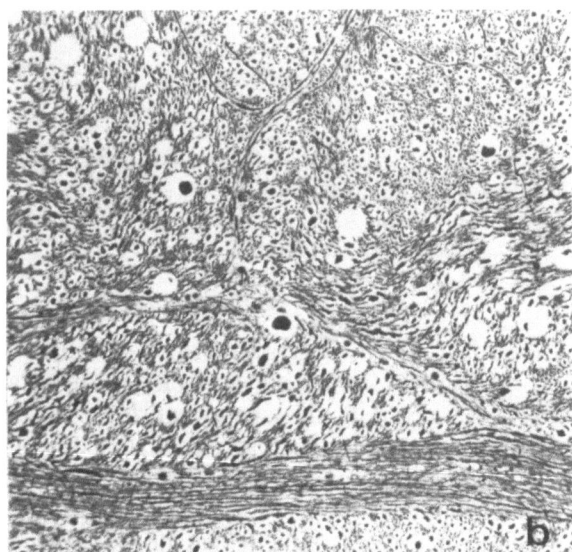

Fig. 7. Diffuse axonal injury. *a* Axonal retraction balls in the pons; patient died 3 days after a head injury; *b* axonal retraction balls in the pons of a subhuman primate subjected to the Penn. 2 device and sacrificed 24 hours later. Both Palmgren × 100 (both from Adams *et al.*[2])

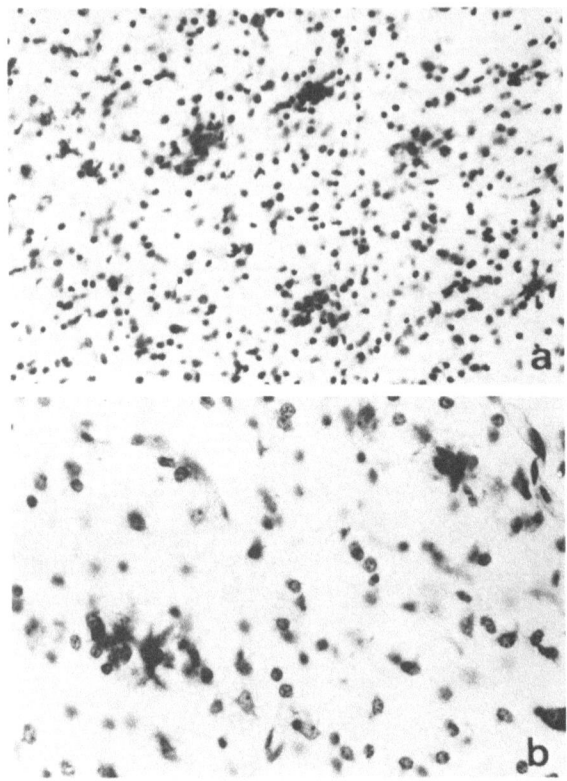

Fig. 8. Diffuse axonal injury. Clusters of microglia in the white matter of the cerebral hemispheres. *a* Patient died 6 weeks after a head injury; *b* subhuman primate subjected to the Penn. 2 device and sacrificed 7 days later. Cresyl violet, *a* × 100, *b* × 200 (*b* from Gennarelli *et al.*[12])

of hypoxic damage in the cerebral cortex: this may take several forms, the commonest being involvement of arterial boundary zones, diffuse damage in the cortex, and involvement of the anterior and middle cerebral arterial territories. The incidence and distribution of hypoxic damage in the cortex of patients with and without DAI is similar (Table 1). Hypoxic damage comprising multiple well-defined foci of neuronal necrosis throughout the cortex was seen in only 4 of the subhuman primates subjected to angular acceleration with the Penn. 1 device: this, however, is not surprising since the animals were carefully monitored so that systemic hypoxia and hypotension did not occur. Such lesions were

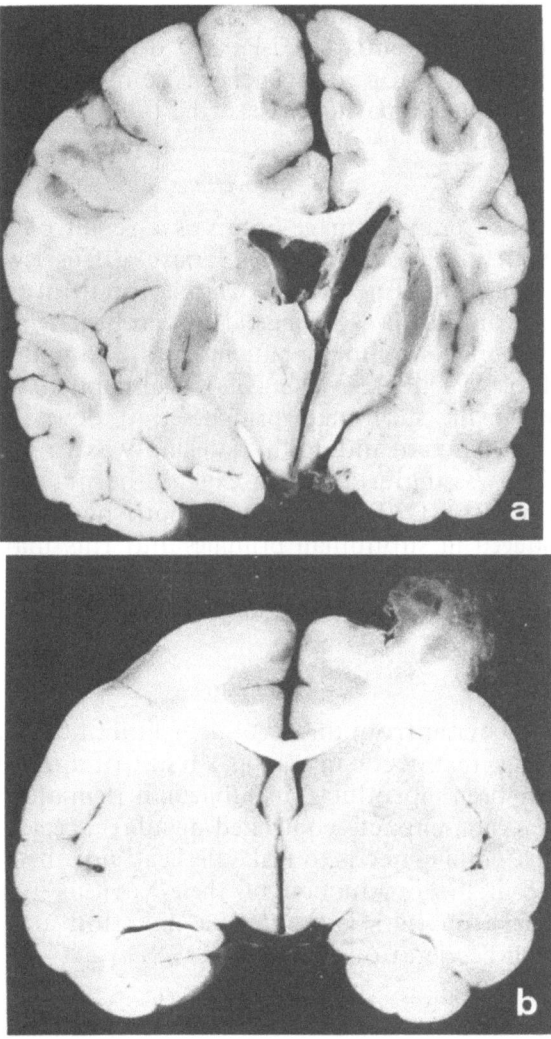

Fig. 9. Unilateral brain swelling. *a* There is diffuse swelling of the left cerebral hemisphere and a shift of the midline structures to the right; patient died 10 days after evacuation of a traumatic SDH, *b* large haemorrhagic external hernia cerebri in the posterior frontal region of the right cerebral hemisphere in a subhuman primate subjected to the Penn. 1 device. A right SDH was evacuated through a craniotomy 47 minutes after injury (*a* from Adams *et al.*[7]; *b* from Adams *et al.*[5])

not seen in the Penn. 2 animals. It is, however, interesting to note that with both the Penn. 1 and Penn. 2 devices there was a high incidence of focal lesions of ischaemic type in the Ammon's horns (Fig. 4 b). The precise pathogenesis of this type of brain damage has not been identified.

Brain Swelling

This type of brain damage occurs as a result of a non-missile head injury in man in two principal forms—diffuse swelling of one cerebral hemisphere (Fig. 9 a), often in relation to an acute subdural haematoma, and diffuse swelling of both cerebral hemispheres. The incidence of brain swelling in patients with and without DAI was similar (Table 1). Diffuse swelling of one hemisphere was a frequent occurrence in the subhuman primates that developed an acute subdural haematoma: indeed this swelling was so severe that if a craniectomy was undertaken, an external hernia developed very rapidly (Fig. 9 b). Diffuse swelling of both hemispheres has not been produced in subhuman primates and this may be at least partly attributable to the fact that in man this type of swelling tends to occur principally in children and adolescents.

Discussion

It should be clear from this account that all of the major types of brain damage that occur in man as a result of a non-missile head injury have been reproduced in subhuman primates subjected to inertial, *i.e.* non-impact, controlled angular acceleration of the head. Thus, nothing needs to strike the head nor the head to strike something in the production of these various types of brain damage. What matters is the degree, direction and duration of acceleration/deceleration impulses.

Acknowledgement

We wish to acknowledge with thanks the help of the Department of Medical Illustration, Southern General Hospital, Glasgow.

    This work was supported by a grant from the Natural Institute of Neurological and Communicative Disorders and Stroke (NS 08803-103).

References

1. Adams, J. H., Graham, D. I., The relationship between ventricular fluid pressure and the neuropathology of raised intracranial pressure. Neuropathol. Appl. Neurobiol. *2* (1976), 323—332.

2. Adams, J. H., Gennarelli, T. A., Graham, D. I., Brain damage in non-missile head injury. In: Recent advances in neuropathology, Vol. 2 (Smith, W. T., Cavanagh, J. B., eds.), pp. 165—190. Edinburgh-London-Melbourne-New York: Churchill Livingstone. 1983.

3. Adams, J. H., Graham, D. I., Diffuse brain damage in non-missile head injury. In: Recent advances in histopathology, Vol. 12 (Anthony, P. P., MacSween, R. N. M., eds.). Edinburgh-London-Melbourne-New York: Churchill Livingstone. In press.

4. Adams, J. H., Graham, D. I., Gennarelli, T. A., Acceleration induced head injury in the monkey. II. Neuropathology. Acta Neuropathol. (Berl.), Suppl. VII (1981), 26—28.

5. Adams, J. H., Graham, D. I., Gennarelli, T. A., Neuropathology of acceleration-induced head injury in the subhuman primate. In: Head injury: Basic and clinical aspects (Grossman, R. G., Gildenberg, P. L., eds.), pp. 141—150. New York: Raven Press. 1982.

6. Adams, J. H., Graham, D. I., Murray, L. S., Scott, G., Diffuse axonal injury due to non-missile head injury in humans: an analysis of 45 cases. Ann. Neurol. 12 (1982), 557—563.

7. Adams, J. H., Graham, D. I., Scott, G., Parker, L. S., Doyle, D., Brain damage in fatal non-missile head injury. J. Clin. Pathol. 33 (1980), 1132—1145.

8. Adams, J. H., Mitchell, D. E., Graham, D. I., Doyle, D., Diffuse brain damage of immediate impact type. Its relationship to "primary brain stem damage" in head injury. Brain 100 (1977), 489—502.

9. Adams, J. H., Scott, G., Parker, L. S., Graham, D. I., Doyle, D., The contusion index: a quantitative approach to cerebral contusions in head injury. Neuropathol. Appl. Neurobiol. 6 (1980), 319—324.

10. Gennarelli, T. A., Head injury in man and experimental animals: clinical aspects. Acta Neurochir. (Wien), Suppl. 32 (1983), 1—13.

11. Gennarelli, T. A., Spielman, G. M., Langfitt, T. W., et al., Influence of the type of intracranial lesion on outcome from severe head injury: a multicenter study using a new classification system. J. Neurosurg. 56 (1982), 26—32.

12. Gennarelli, T. A., Thibault, L. E., Adams, J. H., Graham, D. I., Thompson, C. J., Marcincin, R. P., Diffuse axonal injury and traumatic coma in the primate. Ann. Neurol. 12 (1982), 564—574.

13. Graham, D. I., Adams, J. H., Doyle, D., Ischaemic brain damage in fatal non-missile head injuries. J. Neurol. Sci. 39 (1978), 213—234.

14. Graham, D. I., McLellan, D., Adams, J. H., Doyle, D., Kerr, A., Murray, L. S., The neuropathology of the vegetative state and severe disability after non-missile head injury. Acta Neurochir. (Wien), Suppl. 32 (1983), 65—67.

15. Peerless, S. J., Rewcastle, N. B., Shear injuries of the brain. Can. Med. Assoc. J. 96 (1967), 577—582.

16. Scott, G., Kerr, A., Murray, L. S., A computerized data retrieval system for brain damage in fatal non-missile head injury. Acta Neurochir. (Wien), Suppl. 32 (1983), 95—97.

17. Strich, S. J., Diffuse degeneration of the cerebral white matter in severe dementia following head injury. J. Neurol. Neurosurg. Psychiat. *19* (1956), 163—185.
18. Strich, S. J., Shearing of nerve fibres as a cause of brain damage due to head injury. Lancet 2 (1961), 443—448.
19. Zimmerman, R. A., Bilaniuk, L. T., Genarelli, T. A., Computerised tomography of shearing injuries of the cerebral white matter. Radiology *127* (1978), 393—396.

Authors' addresses: J. H. Adams, Professor of Neuropathology, D. I. Graham, Professor of Neuropathology, Department of Neuropathology, Institute of Neurological Sciences, Southern General Hospital, Glasgow, G 51 4 TF, Scotland; Dr. T. A. Gennarelli, Associate Professor of Neurosurgery, Division of Neurosurgery, University of Pennsylvania, 3400 Spruce Street, Philadelphia, PA 19104, U.S.A.

Acta Neurochirurgica, Suppl. 32, 31—53 (1983)

*Guys Hospital Medical School and ** St. George's Hospital Medical School, London, *** University of Oxford, and **** University of Birmingham, U.K.

Deposition of Scar Tissue in the Central Nervous System

By

M. Berry*, W. L. Maxwell**, A. Logan****, A. Mathewson*, P. McConnell***, Doreen E. Ashhurst**, and G. H. Thomas****

With 9 Figures

Summary

Standard parasagittal lesions were placed stereotactically in the cerebral hemispheres of neonatal and adult rats in order to compare scarring in the immature and mature animal. Lesions were examined by light and electron-microscopy and immunofluorescence to study the astrocyte reaction, collagen deposition, and the formation of the basement memebrane of the glia limitans.

Normal mature scarring characterized by the deposition of collagen, astrocyte end-feet alignment over a glia limitans, and the permanent presence of mesodermal cells (fibroblasts and macrophages) in the core of the lesion, does not occur in wounds before 8–10 days post-partum (dpp). Instead there is no deposition of collagen, and only a transitory astrocyte response occurs with the formation of an interrupted glia limitans. These latter features disappear with time so that the wound is ultimately obliterated by the growth of axons and dendrites through the lesion. Mature scarring is attained over 8–12 dpp when increasing amounts of collagen are deposited and a continuous permanent glia limitans is formed.

The acquisition of the mature response to injury from 8–12 dpp may be correlated with the presence of increasing titres of a fibroblast growth factor (FGF), derived from autolytic digestion of injured brain tissue. We have investigated FGF activity using a 3 T 3 fibroblast tissue culture assay to detect mitogenic activity in brain extracts from rats lesioned at different ages and from leukodystrophic mice which have no myelin.

Our results show that high titres of FGF are present in the developing brain long before myelination commences, and that normal levels of FGF are found in the brains of leukodystrophic mice which have no myelin. Scarring in brain lesions in these mutants is quite normal.

Keywords: Brain lesions; mesodermal/glial scarring; maturation of scar tissue; brain derived FGF; leucodystrophic mutant mice; shiverer; myelin deficient; quaking.

Introduction

Both glia and mesodermal cells participate in the formation of scar tissue in the central nervous system (CNS). The mesodermal elements include haematogenous cells[1,24] and fibroblasts[38] derived from the meninges and other connective tissue sources[27]. Astrocytes constitute the glia component exclusively[7,12,13,29,33,44,45,51,52] and probably interact with mesoderm to produce basement membrane[2,36,38,40,44]. The scar that eventually forms across a breach in the blood-brain barrier essentially reconstitutes the glia limitans externa[46,49], a trilaminate structure composed sequentially of astrocytic end-feet, basement membrane, and collagen fibrils and fibroblasts from within outwards. Interestingly, this response is only a feature of adult brain lesions. In very young animals little scar tissue is deposited in wounds in the CNS[5,49]. Indeed, mature scarring, involving both an astrocytic reaction and the influx of fibroblasts, appears only in the late neonatal period in rats. Before this time, the tissue simply grows together after injury without the intervention of a glial/collagen cicatrix. A study of the development of scarring in the brain might therefore help to elucidate many of the enigmas that presently cloud an understanding of the injury response of the CNS, like, for example, the nature of the stimuli which initiate collagen and basement membrane synthesis, mobilize the astrocytic reaction, and attract mesodermal cells into the wound.

Trophic factors are secreted by macrophages in the wound[9,22,33,34,53] and include endothelial and fibroblast mitogens and collagen synthesis activators[9,22]. A factor has recently been discovered within the brain which is also mitogenic for fibroblasts[17-20]. However, there is much conjecture relating to both the origin of this brain fibroblast growth factor (FGF) and to the role of FGF in brain scarring. For example, it has been suggested that FGF is derived from myelin basic protein (MBP)[20,55], while others maintain that FGF is not myelin derived[10,26,51] but probably originates from an acid protein of unknown origin[51]. Certainly, the proposition that FGF is myelin derived correlates with the observation that the appearance of scarring is coincident with the onset of myelination[16,39,41]. We have, therefore, studied scarring in neonatal and adult rat brain, and in the CNS of leucodystrophic mutants (shiverer, myelin deficient [mld] and quaking) using electron-microscopy, and immunocytochemical techniques for different collagen types and for astrocytes. FGF titres have also been estimated and correlated both with the amounts of collagenous material deposited in wounds and with the state of myelination.

Materials and Methods

1. Light Microscopy

a) Animals

Rats were lesioned at 2, 4, 6, 8, 10, 12, 18 and 38 days *post-partum* (dpp) using a stereotactic instrument. The lesion was made with an iridectomy knife to a depth of 3.5 mm along a 4.5 mm line parallel with the sagittal suture, 3 mm lateral to the mid-line and spanning the fronto-parietal suture (Fig. 1). All animals were killed at 10 days post-lesioning (dpl).

Adult normal and leucodystrophic mutant mice (quaking, shiverer and mld) were similarly lesioned (using appropriately scaled down co-ordinates) and groups of 5 animals allowed to survive until 20 dpl.

b) Histochemistry

The presence of carbohydrate ("vic"-glycol) groups in wounds was detected with the light microscope using the periodic acid silver methenamine technique in both rats and mice in groups of 5 animals at each age, 10 dpl and 5 dpl. Such compounds are probably components of the matrix of a scar in the brain.

2. Electron-Microscope Studies

a) Animals

(i) Adult Rats. The right cerebral hemisphere of Wistar rats was stereotactically lesioned (Fig. 1) at 30 dpp and the animals allowed to recover for 1, 2, 4, 8, 16, 30 and 60 dpl before they were killed.

(ii) Neonatal Animals. Similar lesions (Fig. 1), using appropriately scaled down co-ordinates, were placed in the brains of neonatal rats aged 2, 4, 8 and 12 dpp and the animals were allowed to survive for 8 dpl.

(iii) Leucodystrophic Mutants. Similar stereotactic lesions (Fig. 1) were also placed in the brains of adult mutant leucodystrophic mice aged 40 dpp and animals were allowed to survive for 20 dpl.

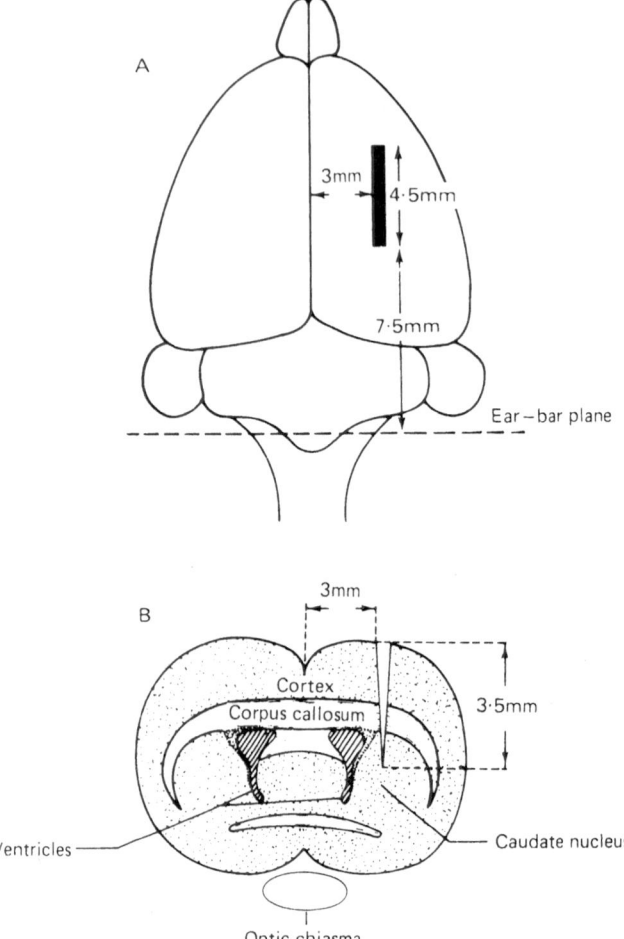

Fig. 1. Diagram of a rat brain showing the dimensions of the knife wound and the structures damaged along the lines of the lesion: *A* position of the lesion on the surface of the hemisphere; *B* position of the lesion in a coronal section of the hemisphere

b) Fixation

All animals were perfused with 5% glutaraldehyde in 0.1 M phosphate buffer 20 g/l sucrose added to give an osmolality of 650 mOsm. The brains were immersed in fixative for 3 hours at 4 °C and then washed overnight in buffer plus sucrose at 4 °C. The region of brain containing the lesion was then cut into 1.5 mm coronal slices and post-fixed for 1 hour in 1% osmium tetroxide in buffer plus sucrose and embedded, through propylene oxide, into TAAB resin.

Silver-gold sections were stained with uranyl acetate and lead citrate and examined with a Philips 201 or 301 electron microscope.

3. Immunocytochemistry

a) Animals

Parasagittal stereotactic lesions (Fig. 1) were made in adult and neonatal rats. Adult rats, aged 30 dpp, were allowed to survive for 4, 8, 16 and 30 dpl. Neonatal rats aged 2, 4, 8 and 12 dpp all survived until 8 dpl.

Animals were killed with chloroform and their brains placed in liquid nitrogen. Frozen sections were cut at 8 μm and air dried on to slides at room temperature.

b) Procedure

Antibodies to types I, III, IV and V collagen were prepared as described by Duance et al.[14]. Incubation with antisera was carried out overnight at 20 °C. After washing for 1 hour, sections were dried and exposed to fluorescein-conjugated anti-rabbit or anti-goat IgG (Wellcome) for 1 hour at 20 °C. After washing again for 1 hour, sections were mounted in gelatin.

4. Fibroblast Growth Factor (FGF)

a) Extraction

Fresh brains were homogenized at 4 °C for 1 minute with 0.15 M $(NH_4)_2 SO_4$ (1.5 W/V) and left for 2 hours, at pH 4.5. After centrifugation (23,000 G for 1 hour), the supernatant was dialysed overnight against 10 vols of deionized water. After a further centrifugation (23,000 G for 1 hour), the supernatant was freeze dried and the protein concentration determined by the method of Lowry[31].

b) Assay

Following the methods of Gospodarowicz[17] we have investigated brain FGF activity in the extracts using a 3 T 3 fibroblast tissue culture assay to detect mitogenic activity in the brains of rats, at different ages, and in those of mature normal mice and murine mutants with severe CNS myelin deficiencies.

Five thousand 3 T 3 cells were seeded into each well of a 96 well Microtitre plate (Sterilin Ltd., Middlesex) in 200 μl of RPM 1 1640 medium with Hepes supplemented with 5% foetal calf serum (fcs), 100 μg/ml penicillin and 100 μg/ml streptomycin (all from Flow Laboratories, Scotland). The plates were incubated at 37 °C in humidified 5% CO_2. The medium was replaced after 8 hours, with 200 μl of the same medium with 0.2% fcs and 1 μg/ml Dexamethasone (Sigma London Chemical Co. Ltd.). Cells were allowed to become quiescent for 36 hours after which time various concentrations of brain extract to be tested, dissolved in medium containing 0.5% crystalline bovine albumin (Sigma London Chemical Co. Ltd.) to minimize non-specific absorption, were added in volumes ranging from 5–15 μl per well. After 12 hours, 10 μl of medium containing 1 μ Ci of (methyl-³H)-thymidine (2 Ci/mmol, Radiochemical Centre, Amersham) and 600 ng of unlabelled thymidine (Sigma London Chemical Co. Ltd.) was added to each well. After 24 hours incubation at 37 °C, the cells were washed twice with 200 μl RPM 1 1640. Each well was swabbed with a cotton-wool bud (Johnsons and Johnsons Ltd., Slough) after adding 50 μl phosphate-buffered saline. Phase examination of wells was used to check that no cells remained after swabbing. The plastic ends of the buds were stuck into the corresponding plasticine filled wells of another Microtitre plate. This second plate was inverted over a glass dish containing 10% cold trichloroacetic acid (TCA), so that the tip of each bud was immersed in the solution. The TCA precipitates the DNA which becomes entrapped in the cotton fibres of the bud. The buds were kept in 10% TCA for 20 minutes. They were then washed in a similar fashion in 5% TCA and washed again for 15 minutes in 95% ethanol. Afterwards they were dried at 37 °C. The cotton tip of each bud was cut and placed in a mini-scintillation vial containing 4 Fisofluor "1" scintillation fluid (Fisons Scientific Apparatus, Leics). The vials were then counted in a liquid scintillation counter (Nuclear, Chicago Mark II). The incorporation of radioactivity into wells with no added extract was taken as the control.

c) Animals

(i) Developmental Study on Rats. Groups of 20 rats were killed at 15, 17, 20 and 22 days post conception (dpc) and groups of 10–15 rats were killed at 8, 10, 12, 20, 30 and 200 dpp. Brains in each group were homogenized and extracts tested for FGF activity as outlined above. Five estimates of DPM/well were made for the brain extracts obtained from animals at each age and standard deviations calculated.

(ii) FGF Activity in Normal and Leucodystrophic Mice. Brain extracts were made from groups of 10 normal mice, shiverer, quaking and mld mutants, and FGF activity was estimated in each case using the technique outlined above. Five estimates of DPM/well were made for each class of animal and standard deviations estimated.

5. Immunocytology for Astrocytes

a) Animals

(i) Mature rats aged 40 dpp received stereotactic lesions as described in Fig. 1. Groups of 5 animals were killed 2, 4, 8, 12, 20 and 30 dpl.

(ii) Immature rats aged 2, 5, 8, 10, 15 and 20 dpp were lesioned similarly and allowed to survive for 4, 8, 20 and 40 dpl (Fig. 1).

b) Immunohistochemistry

The brains of all animals were fixed by perfusion with 4% paraformaldehyde in phosphate buffer, immersed in fixative overnight, dehydrated in graded alcohols and embedded in Steedman's[47] polyester wax (B.D.H.). Sections were then cut at 5 μm, mounted and stained by the indirect immuno-fluorescence method or by the PAP method of Sternberger[48] using primary anti-sera against glial fibrillar acid protein (GFAP).

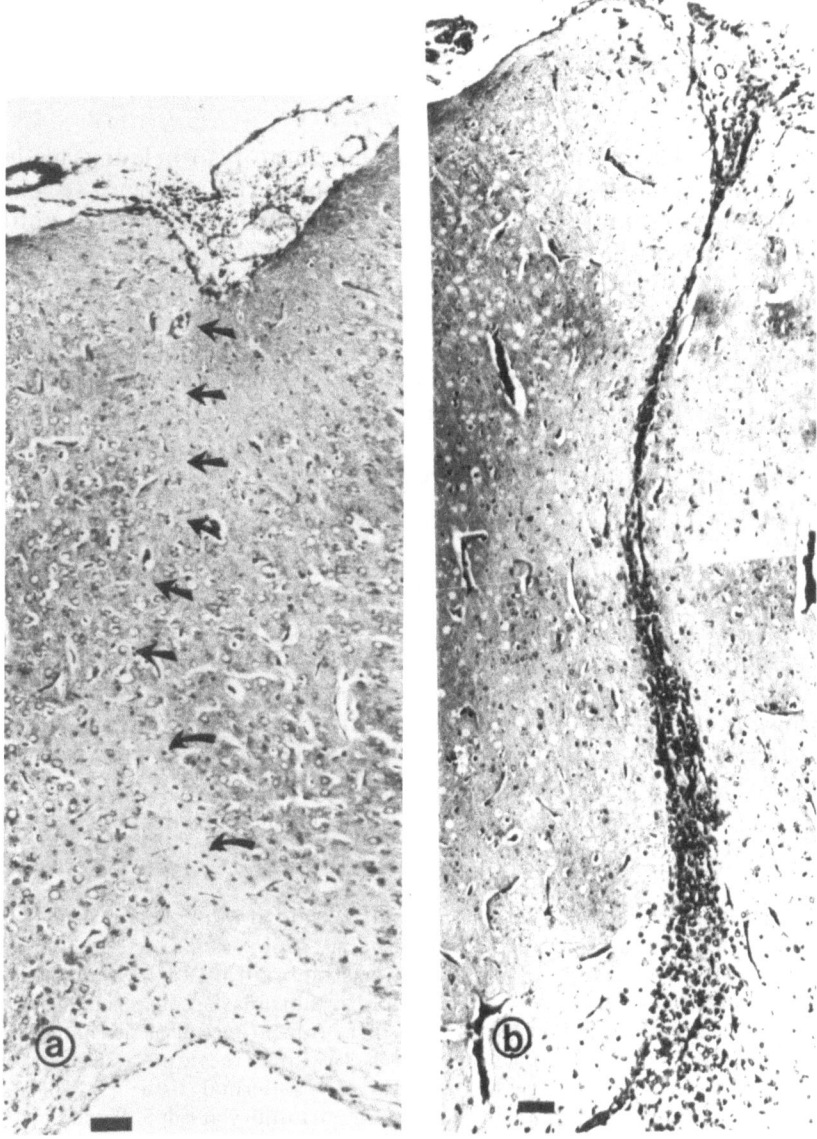

Fig. 2. *a* Lesion in the cerebral cortex of a 30-day-old rat after injury at 2 dpp. The arrows show that the lesion is marked by a pale staining area extending from the ventricle to the pial surface. *b* A lesion in the cerebral cortex 30 dpl made 12 dpp. Note the deposition of collagen and basement membrane material in the lesion (darkly staining material) and the accumulation of fibroblasts and macrophages (representative periodic acid methenamine-silver stained coronal sections, marker = 50 μ)

Results

1. Histochemistry

Stainable carbohydrate material did not appear in lesions made until 8 dpp. Before 8 dpp carbohydrate material was never seen in wounds at 10 dpl; thereafter, increasing amounts appeared until 12 dpp when the lesions were indistinguishable from those of adult animals (Fig. 2). Associated with these changes, increasing numbers of fibroblasts, collagen, basement membrane, and macrophages appeared in brain wounds with age.

The amount of stainable material in all leucodystrophic mutant mice, at 50 dpl, was indistinguishable from that in normal wounds.

2. Electron-Microscopy

a) Mature Response in Rats (Figs. 3 a–c and 4 a and b)

In the adult animal the lesion is initially filled with erythrocytes and by 2 dpl, monocytes and granulocytes appear. Macrophages are most numerous at 4 dpl, and then become progressively less frequent to reach a low level by 16 dpl which is maintained until 60 dpl. The reduction in macrophages from 8 dpl coincides with a rise in the number of cells termed microglia-like which appear either within the scar itself or in the adjacent neuropil. Fibroblasts invade the lesion at 4 dpl and produce a collagenous matrix. They penetrate the lesion depths by 8 dpl. There is then a dramatic loss of

---

Fig. 3. *a* Appearance of a lesion in an adult rat brain at 8 dpl, showing the core of mesodermal cells containing fibroblasts, macrophages and collagen fibrils (× 1,400). *b* Appearance of a lesion in an adult rat brain at 16 dpl. The edge of the neuropil is seen on each side with macrophages, fibroblasts, collagenous material and blood vessels filling the wound (× 1,700). *c* Appearance of a lesion in an adult rat brain at 30 dpl. Neuropil is seen on each side with astrocyte processes lining the wound which is filled with vascularized mesodermal tissue (× 1,700). *d* Appearance of a lesion in a rat brain at 8 dpl after injury at 4 dpp. The line of the lesion is marked by an interrupted cavity but there is no organized astrocyte reaction and an absence of recognizable macrophages, fibroblasts and collagen (× 1,300). *e* Appearance of a lesion in a rat brain at 8 dpl after injury at 4 dpp. The cytoplasm of one definitive macrophage is seen and a number of astrocytes are present within the lesion, along with cells with darkly staining cytoplasm but there is an absence of an organized scar (× 1,200). *f* Appearance of a lesion in a rat brain at 8 dpl after injury at 8 dpp. Astrocytes have become aligned along the lesion edge and a mesenchymal core is beginning to organize containing recognizable macrophages, fibroblasts and collagen fibrils (× 1,700)

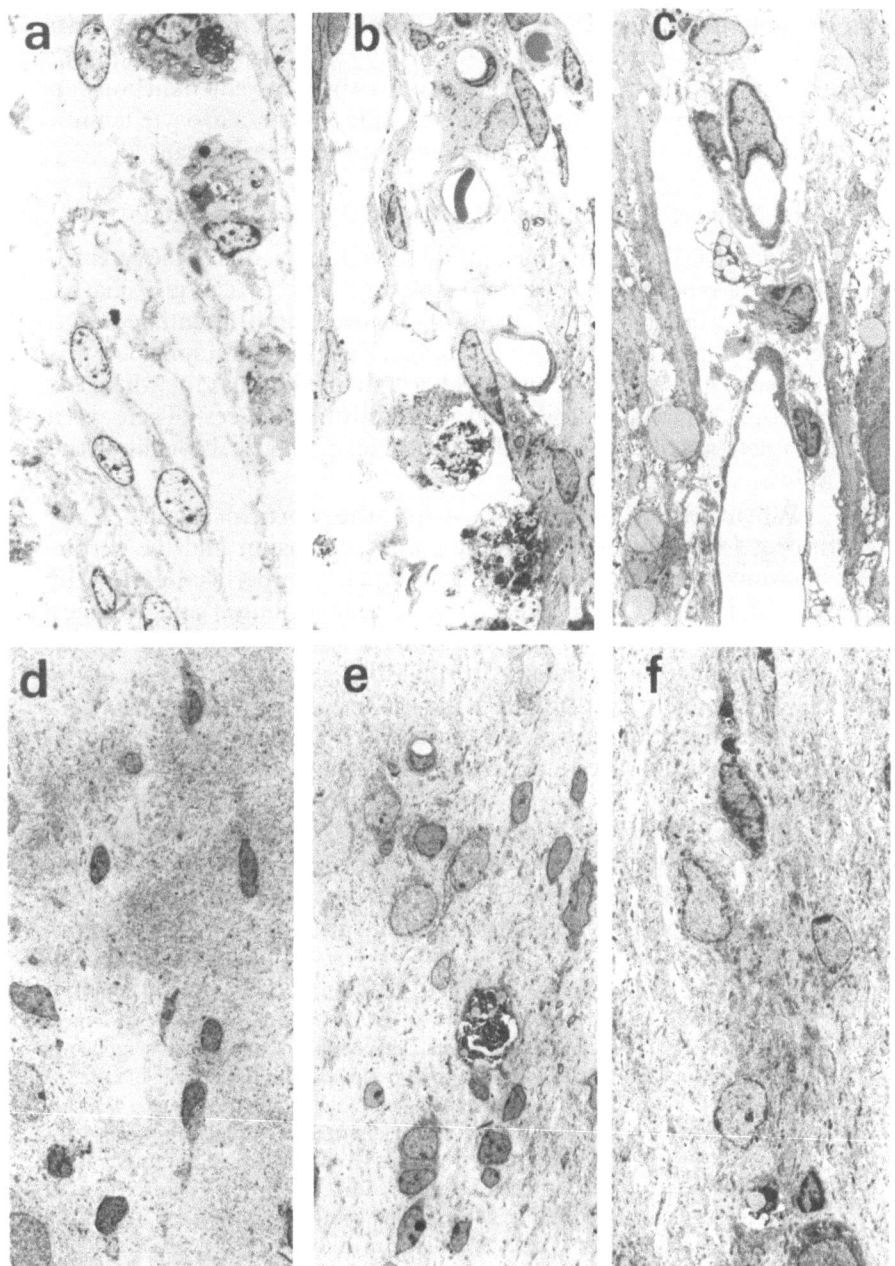

Figs. 3 a–f

fibroblasts with few remaining by 16 dpl. In specimens older than 8 dpl, an astrocyte lamina is formed by a series of overlapping astrocyte processes attached to each other by gap junctions; a basement membrane lies on the scar side of the astrocyte lamina throughout the lesion to form a continuous glia limitans.

b) Neonatal Response in Rats (Figs. 3 d–f and 4 c and d)

Normal mature scarring is not observed in lesions of the cerebral cortex in rats before 8–10 dpp. At 2 dpp the cortical grey matter contains a population of neuroblasts and undifferentiated glial cells; myelin sheaths are not present in the developing corpus callosum. Cells are not concentrated at the lesion margin, where the neuropil exhibits exaggerated extracellular spaces. There is a complete absence of fibroblasts or macrophages throughout the lesion.

When a lesion is made at 4 dpp, the cortical edges become apposed by 8 dpl but, near the corpus callosum and the corpus striatum, the wound remains open. In the superficial cortex the wound is lined by meningeal cells, macrophages and collagen fibrils. Deeper in the cortex, and in the corpus callosum and corpus striatum, only an occasional fibroblast is seen at the margin of the neuropil. Small groups of macrophages are also present at the edges of the neuropil bordering the lumen of the lesion. Collagen fibrils are absent within the lumen and no basement membrane occurs at its edge.

At 8 dpp, there is a deep downgrowth of meninges into the wound where the cortical edges are apposed while, in the corpus callosum and the corpus striatum, there is a large patent lumen. In the area of cortex where the lesion margins are apposed, a small

---

Fig. 4. *a* Appearance of the wound in the brain of an adult rat at 8 dpl. The margin of the cicatrix is lined by astrocytes covered by an incomplete basement membrane. The core of the lesion contains a macrophage, a fibroblast cytoplasmic process and debris (× 9,500). *b* Appearance of the wound in the brain of an adult rat at 30 dpl showing the margin of the scar with a continuous glia limitans comprising astrocytic end-feet, basement membrane, collagen fibrils and fibroblasts (× 10,000). *c* Appearance of the margin of a scar in the brain of a rat at 8 dpl after injury at 4 dpp. There is no organized astrocytic lamina and no basement membrane or collagen. The darkly staining cells are probably microglia (× 6,000). *d* Appearance of the margin of a scar in the brain of a rat at 8 dpl after injury at 8 dpp. A well-defined mesenchymal core is present containing macrophages, fibroblasts and collagen fibrils juxtaposed to a continuous basement membrane overlying astrocytic end-feet (× 8,000)

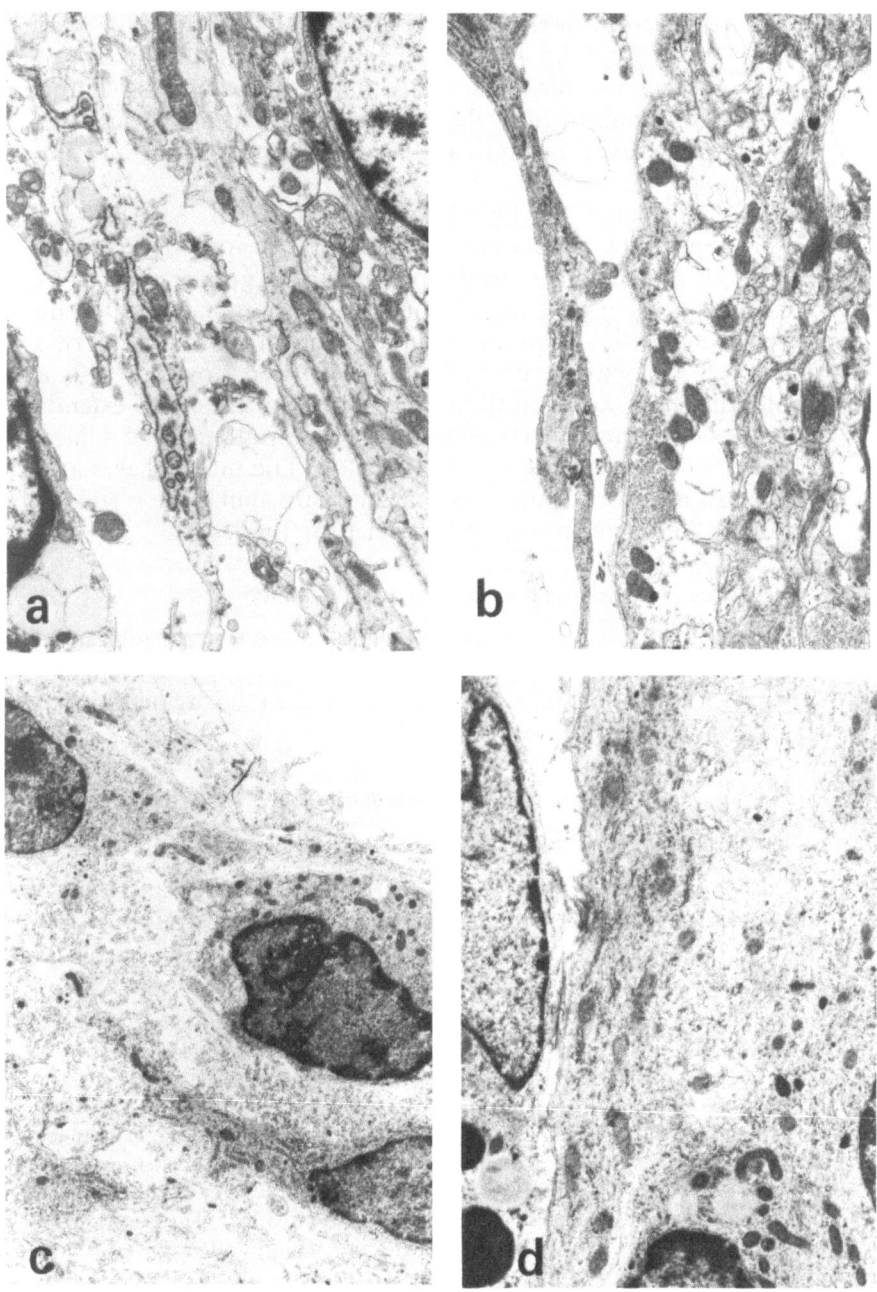

Figs. 4 a–d

number of macrophages/microglia-like cells are present, together with astrocyte processes. A basement membrane is often, but not always, present in this area, but a band of macrophages and astrocytes accumulate along the lesion margin. The astrocytes form an incomplete layer between the macrophage-like cells and the neuropil.

After lesioning at 12 dpp, the cortical edges of the lesion are closely apposed by 8 dpl, and the line of the lesion appears more discrete than earlier, due to the presence of a greater number of macrophages and astrocytes. A thin astrocyte lamina separates the mesenchymal cells from the neuropil and, between these two structures, a basement membrane is deposited, together with a small number of collagen fibrils. The lumen of the lesion extends from the deep cortex into the nucleus caudate-putamen and is lined by a band of macrophages and astrocytes. The macrophages and fibroblasts are more numerous than at 8 dpp, but there is still no organized fibrotic scar within the lumen of the lesion.

c) Adult Normal and Leucodystrophic Mice

When the structure of the scar is compared in normal adult mice and shiverer, quaking and mld murine mutants of the same age (40 dpp) at 20 dpl, no differences are detected in the organization of the scar in any of the animals.

3. *Immunocytochemistry*

a) Mature Response (Figs. 5 a–d)

At 8 dpl, collagen fibrils are present in the extracellular spaces of the scar. Immunocytochemical techniques demonstrate that the fibrils are composed of types I and III collagens. The collagen fibrils

---

Fig. 5. *a–d* are, respectively, immunofluorescence micrographs of type I, III, IV and V collagens at 16 dpl in lesions in the cerebral cortex of adult rats. Note that discrete fibrils are stained with types I and III antibodies whilst staining with antibodies to types IV and V is limited to basement membrane along the margins of the wound and to blood vessels (× 400). *e–h* are, respectively, immunofluorescence micrographs of lesions in rat brains at 8 dpl, after injury at 2 dpp, stained with antibodies to collagen types I, III, IV and V. Antibodies of all types stain the pia/arachnoid membrane and blood vessels in the hemisphere. The lesion edge is, however, unstained (× 100). *i–l* are, respectively, immunofluorescence micrographs of lesions in rat brains at 8 dpl, after injury at 12 dpp, stained with antibodies to collagen types I, III, IV and V. The staining is similar to that seen in Fig. 5 *a–d* but less intense (× 300)

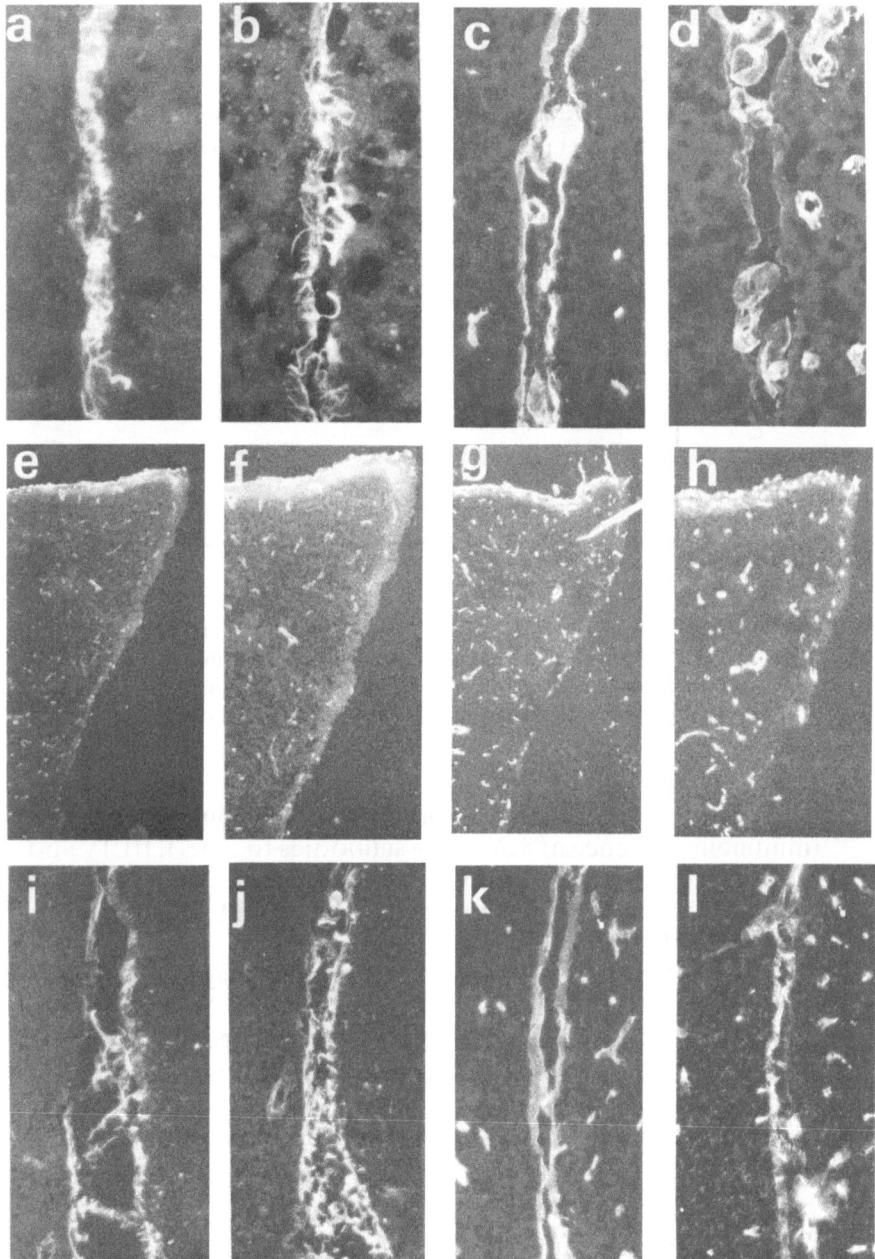

Figs. 5 a–l

are markedly less numerous in the scar nearer the corpus callosum than the grey matter. A basement membrane, along the astrocyte lamina which binds antibodies to types IV and V collagens also appears at 8 dpl; the basement membrane of the blood vessels within both the cicatrix and the neuropil binds the same antibodies.

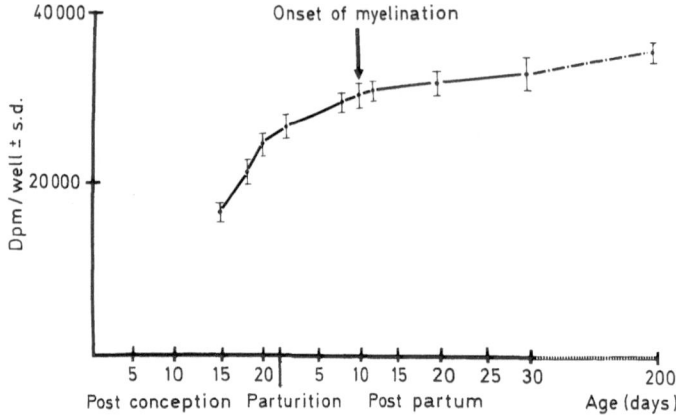

Fig. 6. DNA synthesis in 3 T 3 cells in response to rat brain extracts (10 μg ml⁻¹ protein) taken at various ages. The incorporation of ^{3}H-thymidine into controls with no added extract was 12,479 ± 1425 DPM/well. Five estimates of DPM/well were made for each point and standard deviations calculated

b) Immature Response (Figs. 5 e–l)

After lesioning animals at 4 dpp, there was a complete absence of immunofluorescence at 8 dpl using antibodies to types I, III, IV and V collagens, although types IV and V collagens are associated with the blood vessels of the neuropil and types I and III with fibrous collagen in the meninges.

At 8 dpp, the lesions contain types I and III collagen in the cortex at 8 dpl, while types IV and V collagen also occur but at the lesion margin. There was no basement membrane collagen at the margin or within the lumen of the deep part of the lesion. All types of collagen are present in lesions made at 12 dpp with a distribution indistinguishable from that in the adult at 8 dpl.

4. FGF Activity

a) FGF Activity at Different Ages

The results of this study are shown in Fig. 6. It is clear that significant values of FGF already exist in the brain at 15 dpc and that levels reach approximate adult values at 10 dpp, at a time when myelination is about to commence. Thus, FGF titres in the brain are not correlated either with the onset of myelination or with the development of scar tissue in brain lesions.

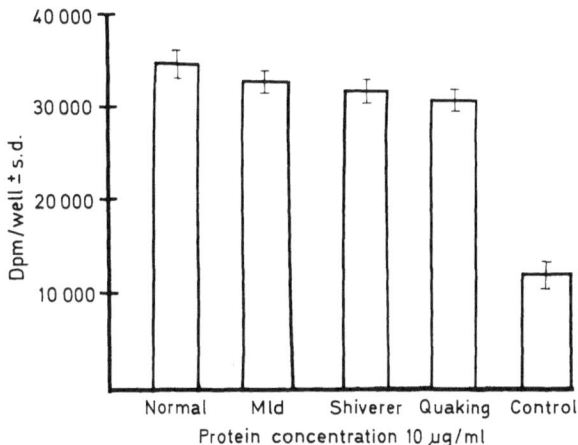

Fig. 7. DNA synthesis in 3 T 3 cells maintained in the presence of extracts from normal and mutant mouse brains. Five estimates of DPM/well were made for each treatment and standard deviations calculated. A 2-way analysis of variance revealed no significant differences in FGF activity in brain extracts between mutant and normal mice. Control values were provided by cultures containing 0.2% fcs but no extract

b) FGF Activity in Normal and Leucodystrophic Mice

The titres of FGF activity extracted from adult normal mice and from mld, shiverer and quaking mice of similar age are not significantly different (Fig. 7).

5. The Response of Astrocytes to Injury

a) Adult Response

By 2 dpl astrocytes around the wound have become GFAP-positive. Their numbers increase with time and reactive astrocytes are seen in areas of cortex increasingly distant from the environs of the lesion. By 8 dpl, the entire cortex contains a uniform density of GFAP-positive astrocytes. The number of reactive astrocytes

M. Berry *et al.*:

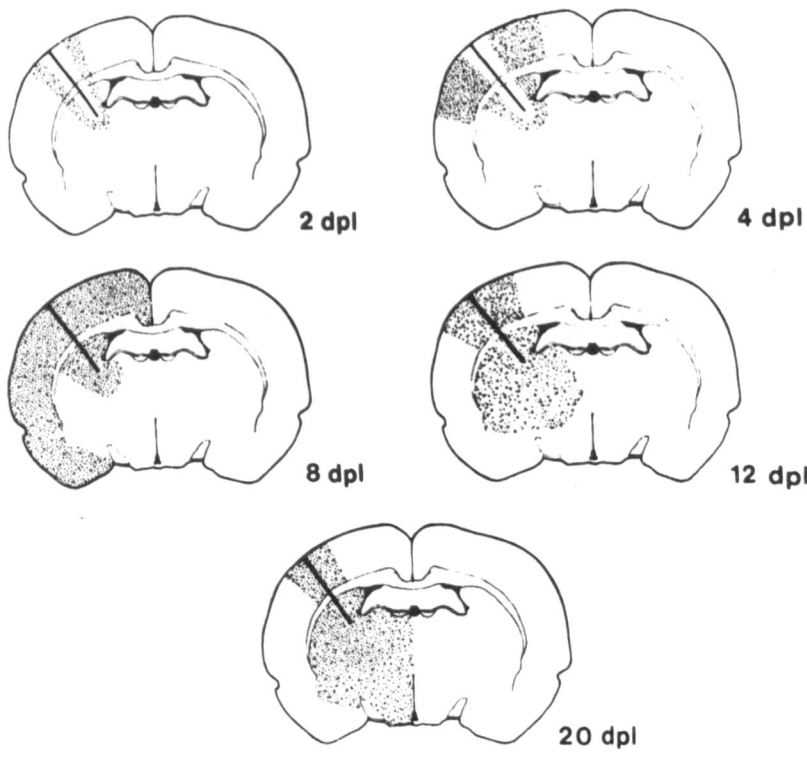

Fig. 8. Diagrammatic representation of the spatial and temporal changes in GFAP astrocyte reactivity (stippled area) after lesioning one hemisphere of adult rats. The cortical response is maximal at 8 dpl; the deep response is maximal at 20 dpl; note the confinement of reactivity to one hemisphere

regresses until by 20 dpl only a thin band of GFAP-positive astrocytes surrounds the site of the lesion. A similar spreading reaction is seen in the deep structures of the cerebrum, but the time course is more protracted (Fig. 8).

Astrocytic foot expansions appear along the wound edge at 8 dpl. These form a continuous lining to the wound edge by 12 dpl and, at 20 dpl, form multiple layers in the deeper areas (Fig. 9 a).

b) Neonatal Response

No reactive astrocytes are seen at any time in the wounds of animals lesioned at 2 dpp. After lesioning at 5 dpp, a small number

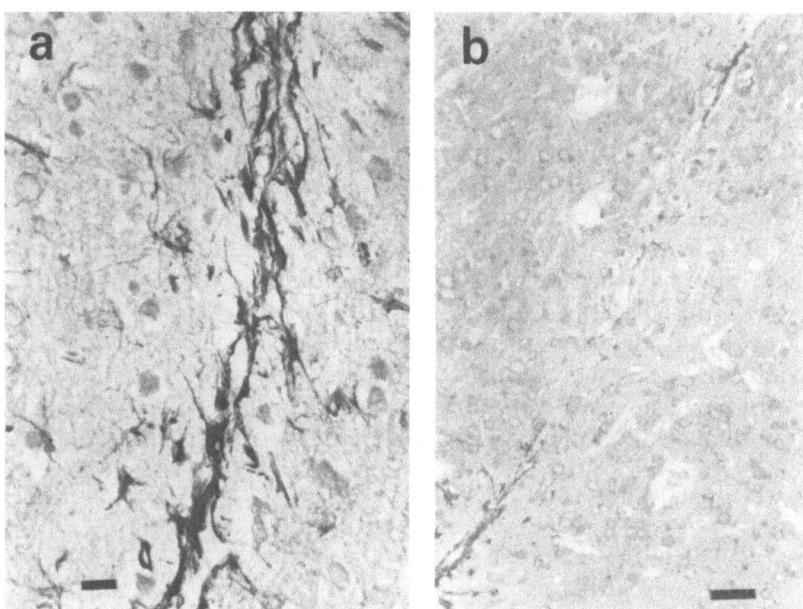

Fig. 9. *a* Reactive astrocytes in a wound in an adult brain 20 dpl. Note the GFAP reactive astrocytic processes forming a continuous lining to the lesion (marker = 10 μm). *b* Reactive astrocytes in a wound in a brain 20 dpl after injury at 4 dpp. The line of the lesion passes diagonally from bottom left. Note the paucity of the astrocyte GFAP reactivity. Foot-processes form an incomplete glia limitans (marker = 50 μm)

of scattered reactive astrocytes appear at the lesion edge at 4 dpl, but these disappear with time. However, after lesioning at 8 and 10 dpp more reactive astrocytes appear at 4 dpl and their end-feet form a line along the length of the lesion, which persists at least until 40 dpl (Fig. 9 b) although no generalized cerebral reaction occurs. By 15–20 dpp the astrocytic response in lesions is indistinguishable from that of adult animals, but a generalized cerebral reaction is not apparent.

Discussion

1. Time Course of Scarring in the Adult

a) Mesodermal Elements

Up until 4 dpl the acute reaction in brain wounds in adult animals is mediated by haematogenous cells. Thereafter, macrophages, fibroblasts and astrocytes play a significant part in reconstituting the glia limitans. Fibroblasts, possibly derived from surrounding connective tissue[27,38], and a few collagen fibrils first appear in the wound between 4–8 dpl. At first the cells and collagenous matrix in the cicatrix are very loosely packed but, over the period 10–15 dpl, the cicatrix becomes greatly condensed which suggests that scar contraction might occur.

Aside from their role as scavengers, macrophages almost certainly influence scarring by secreting trophic factors which affect both fibroblast and endothelial cell activities[9,22,34,35,54]. The reciprocal relationship between the disappearance of macrophages and the appearance of microglia-like cells over the 4–18 dpl period supports the suggestion of Imamato and Leblond[24] and Adrian and Williams[1] that macrophages transform into microglia-like cells.

b) Glial Cells

(i) Generalized Response

Although a diffuse reaction by astrocytes to injury has been described[6,23], neither the time course of their response nor the curious restriction of reactive astrocytes to the ipsilateral hemisphere have previously been reported. The factors that mediate this diffuse response and the possible function of the reaction are both mysteries.

(ii) Formation of the Glia Limitans

Although reactive astrocytes organize themselves at the wound edge in the early stages of cicatrization, a definitive glia limitans is not found until fibroblasts and collagen fill the wound[40,49]. Thus, basement membrane collagens do not appear in the wound until 8 dpl, but then completely line the lesion by 12 dpl. This possible dependence of epithelial cells on the presence of fibroblasts for the production of basement membrane is seen in other tissues[3,8] and could implicate fibroblasts as organizers of astrocytes for the production of a glia limitans[43].

2. Time Course of Scarring in the Neonate

a) Mesodermal Elements

No fibroblasts appear in brain wounds of animals before 8 dpp. Thus, a scar is not formed after injury at this time. This response is not correlated with an absence of FGF, but could be related to the absence of myelination, or competent macrophages, or both, since degenerating myelin could delay macrophages in the lesion long enough for them to produce a factor that stimulates collagen synthesis in fibroblasts[9,22,34,49,54].

b) Glial Elements

Astrocytes do not become organized to form a glia limitans before 8 dpp. The development of the latter is correlated with the appearance of fibroblasts and collagen in the wound and could indicate that the development of a mature scar depends on the attraction of fibroblasts into the wound area. The latter, in turn, might organize astrocytes to form end-feet at the mesodermal interface and secrete types IV and V collagen to form a basement membrane. Sumi and Hager[49] suggested that the paucity of reaction in the neonate might be caused by astrocytic hypoplasia possibly related to the very rapid removal of necrotic debris by macrophages.

The coincidence of the appearance of scarring with the onset of myelination in the brain could be related to an effect of degenerating myelin or oligodendrocytes on macrophages. Macrophages could stimulate fibroblasts to produce collagen[9,22,34,35,49,54]. The presence of myelin/oligodendrocyte degradation products in the wound might delay macrophages in the lesion long enough for their trophic factors to reach sufficient levels to initiate collagen synthesis. However, at 8 dpp, in rats, a premyelination oligodendrocytosis is underway[4,16,39] and little myelin has formed[25,39,41]. Moreover, in leucodystrophic mutants shiverer, mld and quaking, little or no myelin is present[15,21,32], but scarring is normal.

The absence of scarring in early brain lesions could be related to an immaturity of function of both glia and mesodermal elements. Maturation might be achieved over the 8–12 dpp period when normal scarring is acquired. However, astrocyte-like cells (radial glia) appear competent to produce a glia limitans externa at a much earlier age[11,42,50]. Moreover, subpial end-feet also appear structurally mature at these early times[28,37] and astrocyte (radial glial) processes are GFAP positive[56].

3. FGF

The possibility that FGF is **MBP** derived[20,55] is not supported by our findings of high titres long before myelination commences in the developing brain, and of normal levels in leucodystrophic

M. Berry *et al.*:

mutants which possess little MBP[15,21,32]. Moreover, since near normal FGF titres are present in neonatal brains, which exhibit no collagenous scarring, it is clear that this mitogen may not be as significant in brain healing as was once envisaged[18].

FGF may be ineffective in the absence of a co-factor which only appears in brain wounds from 8 dpp onwards. Alternatively, FGF could be be derived from an acid protein[30,51], but our work sheds no light on its likely source within the brain.

Acknowledgements

We thank Gill Fellows for technical assistance, Kevin Fitzpatrick for the photography and Margaret Collins for preparing the manuscript. The work was supported by the M.R.C. and Nuffield Foundation.

References

1. Adrian, K., Williams, M. G., Cell proliferation in injured spinal cord. An electron microscopic study. J. comp. Neurol. *151* (1973), 1—24.
2. Barrettt, C. P., Guth, L., Donati, E. J., Krikorian J. G., Astroglial reaction in the grey matter of lumbar segments after mid-thoracic transection of the adult rat spinal cord. Exp. Neurol. *13* (1981), 565—577.
3. Bernfield, M. R., Banerjee, S. D., The basal lamina of epithelial-mesenchymal morphogenetic interactions: In: Biology and chemistry of basement membranes (Kefalides, N. A., ed.), pp. 137—148. New York: Academic Press. 1978.
4. Bensted, J. P. M., Dobbing, J., Morgan, R. S., Reld, R. T. W., Payling Wright, G., Neurological development of myelination in the spinal cord of the chick embryo. J. Embryol. Exp. Morphol. *5* (1957), 428—437.
5. Berry, M., Henry, J., Response of neonatal CNS to injury. Neuropath. Appl. Neurobiol. *2* (1975), 166.
6. Bignami, A., Dahl, D., The astroglial response to stabbing. Immunofluorescence studies with antibodies to astrocyte specific protein (GFA) in mammalian and sub-mammalian vertebrates. Neuropath. Appl. Neurobiol.*2* (1976), 99—110.
7. Bignami, A., Ralston, H., The cellular reaction to Wallerian degeneration in the central nervous system of the cat. Brain. Res. *13* (1969), 444—461.
8. Bluemink, J. G., Maurik, P. van Lawson, K. A., Intimate cell contacts at the epithelial/mesenchymal interface in embryonic mouse. J. Ultrastruct. Res. *55* (1976), 257—270.
9. Boutwell, T. K., Factors promoting epidermal cell proliferation. In: The surgical wound (Duneen, P., Hildick-Smith, G., eds.), pp. 90—96. Philadelphia: Lea and Febiger. 1981.
10. Chiang, T. M., Whitaker, J. N., Seyer, J. M., Kang, A. H., Effect of peptides of bovine myelin basic protein in dermal fibroblasts. J. Neurosci. Res. *5* (1980), 439—445.

11. Choi, B. M., Lapham, L. W., Radial glia in the human foetal cerebrum: A combined Golgi, immunofluorescent and electron microscopic study. Brain Res. *148* (1978), 295—311.
12. Colmant, H. J., Allgemeine Histopathologie der Glia. Acta Neuropath. (Berl.) Suppl. IV (1968), 61—76.
13. Cook, R. D., Wisniewski, H. M., The role of oligodendroglia and astroglia in Wallerian degeneration of the optic nerve. Brain Res. *61* (1973), 191—206.
14. Duance, J. C., Restall, D. J., Beard, H., Bourne, F. J., Barley, A. J., The location of three collagen types in skeletal muscle. FEBS Lett. *79* (1977), 248—252.
15. Dupouney, P., Jacque, C., Bourne, J. M., Cesselin, F., Privat, A., Baumann, N., Immunochemical studies on myelin basic protein in shiverer mouse devoid of major dense line of myelin. Neurosci. Lett. *12* (1979), 113—118.
16. Friede, R. L., A histochemical study of DPN-diaphorase in human white matter; with some notes on myelination. J. Neurochem. *8* (1961), 17—30.
17. Gospodarowicz, D., Localisation of a fibroblast growth factor and its effect alone and with hydrocortisone on 3 T 3 cell growth. Nature *249* (1974), 123—127.
18. Gospodarowicz, D., Humoral control of cell proliferation. The role of fibroblast growth factor in regeneration, angiogenesis, wound healing and neoplastic growth. In: Membranes and neoplasia: New approaches and strategies (Marchesi, V. T., ed.), pp. 1—19. New York: Alan R. Liss. 1976.
19. Gospodarowicz, D., Rudland, P., Lindstrom, J., Benirschke, K., Fibroblast growth factor, localization, purification, mode of action and physiological significance. Adv. Metab. Disord. *8* (1975), 301—335.
20. Gospodarowicz, D., Lui, G. -M., Cheng, J., Purification in high yield of brain fibroblast growth factor by preparative isoelectric focusing at pH 9.6. J. Biol. Chem. *257* (1982), 12266—12276.
21. Hogan, E. L., Animal models of genetic disorders of myelin. In: Myelin (Morell, P., ed.), pp. 489—520. New York: Plenum Press. 1977.
22. Hunt, D. K., Andrews, W. S., Halliday, B., Greenburg, G., Knoghton, D., Clark, R. A., Thakrai, K. K., Coagulation and macrophage stimulation of angiogenesis and wound healing. In: The surgical wound (Duneen, P., Hildick-Smith, G., eds.), pp. 1—18. Philadelphia: Lea and Febiger. 1981.
23. Ibrahim, M. Z. M., Glycogen and its related enzymes of metabolism in the central nervous system. Adv. Anat. Embryol. Biol. *52* (1975), 3—89.
24. Imamoto, K., Leblond, C. P., Presence of labelled monocytes, macrophages and microglia in association with a stab wound of the brain after an injection of bone marrow cells labelled with ^{3}H-uridine into rats. J. comp. Neurol. *174* (1977), 255—280.
25. Jacobson, S., Sequence of myelination in the brain of the albino rat. A. Cerebral cortex, thalamus and related structures J. comp. Neurol. *121* (1963), 5—29.
26. Kellet, J. G., Tanaka, T., Rowe, J. M., Shiu, R. P. C., Friesen, H. G., The characterization of growth factor activity in human brain. J. Biol. Chem. *256* (1981), 54—58.

27. Krikorian, J. G., Guth, L., Donati, E. J., Origin of connective tissue scar in the transected rat spinal cord. Exp. Neurol. *72* (1981), 698—707.
28. Landis, D. M. D., Reese, T. S., Arrays of particles in freeze fractured astrocytic membrane. J. Cell. Biol. *60* (1974), 316—328.
29. Latov, N., Nilaver, G., Zimmerman, E. A., Johnson, W. G., Silverman, A.-J., Defendini, R., Cote, L., Fibrillar astrocytes proliferate to brain injury. Dev. Biol. *72* (1979), 381—384.
30. Lemmon, S. K., Riley, M. C., Thomas, K. A., Hoover, G. A., Maciag, T., Bradshaw, R. A., Bovine fibroblast growth factor: comparison of brain and pituitary preparations. J. Cell. Biol. *95* (1982), 162—169.
31. Lowry, O. H., Roosebrough, N. J., Farr, A. L., Randall, R. F., Protein measurements with Folin phenol reagent. J. Biol. Chem. *193* (1951), 265—275.
32. Matthieu, J.-M., Ginaleski, H., Friede, R. L., Cohen, S. R., Low myelin basic protein levels and normal myelin in peripheral nerves of myelin deficient mice (mld). Neuroscience *5* (1980), 2315—2320.
33. Murabe, Y., Ibata, T., Sano, Y., Morphological studies on neuroglia. II. Response of glial cells to kianic acid-induced lesions. Cell Tissue Res. *216* (1981), 569—580.
34. Nathan, C. F., Cohen, Z. A., Cellular components of inflammation, monocytes and macrophages. In: Textbook of rheumatology (Kelley, W. N., Harris, E. D., jr., Ruddy, S., Sledge, C. B., eds.), pp. 136—162. Philadelphia: Saunders.
35. Nathan, C. F., Murray, H. W., Cohn, Z. A., The macrophage as an effector cell. N. Eng. J. Med. *303* (1980), 622—626.
36. Persson, L., Cellular reactions to small cerebral stab wounds in the rat frontal lobe. Virch. Arch. B. Cell. Path. *22* (1976), 21—37.
37. Peters, A., Palay, S. L., Webster, H. de F., The fine structure of the nervous system. Philadelphia: Saunders. 1976.
38. Ross, R., The fibroblast and wound repair. Biol. Rev. *43* (1968), 51—96.
39. Schonbach, J., Hu, J. K., Friede, R. L., Cellular and chemical changes during myelination: Histologic, autoradiographic, histochemical and biochemical data on myelination in the pyramidal tract and corpus callosum of rat. J. comp. Neurol. *134* (1968), 21—38.
40. Schultz, R. L., Pease, D. C., Cicatrix formation in rat cerebral cortex as revealed by electron microscopy. Amer. J. Path. *35* (1959), 1017—1042.
41. Seggie, J., Berry, M., Ontogeny of interhemispheric evoked potentials in the rat: Significance of myelination of the corpus callosum. Exptl. Neurol. *35* (1972), 215—232.
42. Sidman, R. L., Rakic, P., Neuronal migration with special reference to developing human brain: A review. Brain Res. *62* (1973), 1—35.
43. Sievers, J., Mangold, U., Berry, M., Allen, C., Schlossberger, H. G., Experimental studies on cerebellar foliation. I. A qualitative morphological analysis of cerebellar foliation defects after neonatal treatment with 6-OHDA in the rat. J. comp. Neurol. *203* (1981), 751—769.
44. Skoff, R. P., The fine structure of pulse-labelled (^{3}H-thymidine) cells in degenerating rat optic nerve. J. comp. Neurol. *161* (1975), 595—612.

45. Skoff, R. P., Vaughn, J. E., An autoradiographic study of cellular proliferation in degenerating rat optic nerve. J. comp. Neurol. *141* (1971), 133—156.
46. Spatz, H., Über die Vorgänge nach experimenteller Rückenmarksdurchtrennung mit besonderer Berücksichtigung der Unterschiede der Reaktionsweise des reifen und des unreifen Gewebes. In: Histologische und histopathologische Arbeiten über die Großhirnrinde (Nissl, F., Alzheimer, A., eds.), pp. 49—354. Jena: G. Fischer. 1921.
47. Steedman, H. F., A new ribboning embedding medium for histology. Nature *197* (1957), 1345.
48. Sternberger, L. A., In: Immuno-cytochemistry. New York: Wiley. 1979.
49. Sumi, S. M., Hager, H., Electron microscope study of experimental porencephaly. J. Neuropath. Exp. Neurol. *27* (1968), 138.
50. Tennyson, V. M., Electron microscopic study of the developing neuroblast of the dorsal root ganglion of the rabbit embryo. J. comp. Neurol. *124* (1965), 267—318.
51. Thomas, K. A., Riley, M. L., Lemmon, S. K., Baglan, N. C., Bradshaw, R. A., Brain-fibroblast growth factor: nonidentity with myelin basic protein fragments. J. Biol. Chem. *25*, 5 (1980), 5517—5520.
52. Vaughn, J. E., Hinds, P. L., Skoff, R. P., Electron microscopic studies of Wallerian degeneration in the optic nerve of the rat. I. The multipotential glia. J. comp. Neurol. *140* (1970), 175—206.
53. Vaughn, J. E., Pease, D. C., Electron microscopic studies of Wallerian degeneration in rat optic nerves. II. Astrocytes, oligodendrocytes and adventitial cells. J. comp. Neurol. *140* (1970), 207—226.
54. Wahl, S. M., Role of mononuclear cells in wound repair process. In: The surgical wound (Dineen, P., Hildick-Smith, G., eds.), pp. 63—74. Philadelphia: Lea and Febiger. 1981.
55. Westall, F. C., Lennon, V. A., Gospodarowicz, D., Brain-derived fibroblast growth factor: identity with a fragment of the basic protein of myelin. Proc. nat. Acad. Sci. U.S.A. *75* (1978), 4675—4678.
56. Woodhams, P. L., Basco, E., Hajos, F., Csillag, A., Balazs, R., Radial glia in the developing mouse cerebral cortex and hippocampus. Anat. Embryol. *163* (1981), 331—343.

Authors' address: Prof. Dr. M. Berry and Dr. A. Mathewson, Anatomy Department, Guy's Hospital Medical School, London SE 1 9 RT; Dr. W. L. Maxwell and Doreen E. Ashhurst, Department of Anatomy, St. George's Hospital Medical School, London SW 17 ORE; Dr. P. McConnell, Department of Human Anatomy, University of Oxford, South Parks Road, Oxford OX 1 3 QX; Dr. A. Logan and Dr. G. H. Thomas, Department of Anatomy, University of Birmingham, Birmingham B 15 2 TJ, U. K.

Acta Neurochirurgica, Suppl. 32, 55—59 (1983)

National Institute for Medical Research, London, U.K.

Formation of Mossy Fibre Connections Between Hippocampal Transplants and the Brain of Adult Host Rats

By

G. Raisman

Summary

Hippocampal primordia can survive transplantation into adult rat hippocampi. The transplants differentiate and form mossy fibre connections with host pyramidal cells, while mossy fibre projections from the host can innervate transplants.

Keywords: Transplantation; hippocampus; mossy fibres; innervation.

There has been a great deal of interest over the last few years in the capacity of transplants to survive and form connections with adult host brains[8,10,11]. With current techniques, only embryonic or early postnatal donor tissue from the central nervous system can survive[4,16]. Little is known about the mechanism of vascularization of the transplants. Even with strains of rat not specially inbred for histocompatibility, however, there is little evidence of immunological rejection of central nervous tissue transplanted into the brain[12].

We have transplanted hippocampal primordia into the hippocampal region of adult rat hosts[12,13]. Donor animals aged from the eighteenth day of embryonic life (E 18) to the first day of postnatal life (P 1) were used. At the time of transplantation some of the hippocampal neurons—the pyramidal cells—had already been formed, while others—the dentate granule neurons and most of the astrocytes—are still not formed from their precursor cells[1, 14, 15].

The survival period was around one month. Most of the experiments were carried out between rats from a locally inbred histocompatible strain (PVG), although equally good results were obtained from Wistar rats. The transplants derived from the earlier age donors were large and had a well organized hippocampal structure. Tritiated thymidine administered to the host after transplantation showed that the dentate granule cells and the astrocytes were largely formed (by mitosis of stem cells in the transplant) after transplantation, whereas the pyramidal cells were unlabelled. Thus, cell division is neither essential for, nor inimical to the survival of transplanted cells. With donor tissues from the later ages, and particularly with P 1 donors, the resulting transplants were much smaller and more fragmented. Unexpectedly, the most effective connections were formed by these later transplants, and the subsequent material refers to the P 1 series. It seems that the fragmentary survival of parts of the transplanted tissue is in some way conducive to the formation of fibre connections to and from the host.

We decided to investigate the regulation of the hippocampal dentate granule cell projections using transplantation. The axons of the dentate granule cells (called mossy fibres) can be readily stained in light microscopic sections by the Timm stain, which is based on their high content of zinc[6, 7, 17]. In the normal rat hippocampus, the dentate granule cells project mossy fibres to the giant pyramidal neurons of field CA 3, but not to the small pyramids of field CA 1[2, 3]. Within CA 3, the mossy fibres terminate on those parts of the dendrites closest to the cell bodies, and especially on the apical dendritic side of these cells. We have used transplants to investigate what factors may determine this pattern of connections. For this purpose, hippocampal primordia were transplanted in and around the junction of fields CA 1 and CA 3 of adult rat host hippocampi.

The material shows both that the transplant can send mossy fibres into the host, and also that the host can project mossy fibres into the transplant. These findings refer to light microscopic material, but a preliminary series of observations with the electron microscope confirms that the transplant to host projection, at least, forms fully differentiated mossy fibre terminals which make synaptic contacts on complex dendritic spines of the host pyramidal neurons.

One of the principal findings of this study was that transplant dentate granule neurons can send mossy fibres into the host field CA 1. This, therefore, is a type of projection not normally present in the hippocampus. Nonetheless, the transplant projection to the host CA 1 is to a highly restricted area—it is juxtacellular, but to the basal aspect of the dendrites, and it extends for no more than 1 mm from the contact zone. This suggests that this projection is based on some form of specific recognition between the transplant axon and the target region.

From a large series of experiments it is clear that for the projection to form it is necessary for the configuration at the interface to fulfil highly stringent conditions. It is essential for the transplant to be in direct contact with the host field CA 1, and for there to be no substantial interposition of neuroglial cell bodies. Even under these circumstances connections will not form unless the axonal side of the transplant dentate granule cells is facing the host. Where the dendritic side of the transplant dentate granule cells (the stratum moleculare) is facing the host, the transplant mossy fibres, even though they invade the inner parts of the dendritic layer of the transplant dentate gyrus, will not cross into the host brain. A final condition is that in the region of the interface there must be no giant (CA 3-type) pyramidal cells in the transplant. If such transplant pyramids are present they attract the transplant mossy fibres to them, and appear to satisfy their growth potential and thus prevent them reaching the host.

It is not clear why the transplant mossy fibres form this abnormal projection to the host field CA 1. Possibly a specific type of deafferentation is necessary, and has been caused by some aspect of the surgical intervention involved in transplantation. Certainly the same layer of the host field CA 1 that receives the transplant mossy fibre projection is normally in receipt of mossy fibre projections in other species, such as cats[9] and hedgehogs[5]. It seems likely that in the rat the CA 1 pyramids may have a latent ability to receive mossy fibres, and this is "revealed" by transplanting a source of mossy fibres into their vicinity.

Mossy fibre projections from the host can also innervate transplants. For them to be demonstrable by the Timm stain it is necessary that the transplant does not have any dentate granule cells of its own in the vicinity (or else the transplant's own mossy fibres will obscure those from the host). The circumstances in which the host will project mossy fibres to the transplant require that the transplant be placed directly in the host mossy fibre pathway. Very fine mossy fibres can be seen entering the transplant, and directing themselves towards transplant CA 3 pyramids. Around the pyramids they form very large granules—comparable to the normal mossy fibre terminals in field CA 3 of the host. In the case where the transplant also contains clumps of smaller, CA 1-type pyramids, the mossy fibres do not form any obvious terminals around them. In fact, some cases show that the mossy fibres may deflect themselves in their course, so as to circumvent CA 1 pyramidal clusters on their way to CA 3 clusters.

In conclusion, we have confirmed that hippocampal primordia can survive transplantation into adult rat hippocampi. The transplants differentiate and form mossy fibre connections (see also [17]). The transplant can give mossy fibres to the host brain, and receive mossy fibres from it. The terminal patterns formed by these mossy fibres appear to be specific to the target regions involved, and bear some relation to normal mossy fibre patterns. The functional status of these connections remains for investigation.

References

1. Bayer, S. A., Development of the hippocampal region in the rat. I. Neurogenesis examined with [³H]-thymidine autoradiography. J. comp. Neurol. *190* (1980), 87—114.
2. Blackstad, T. W., Kjaerheim, A., Special axo-dendritic synapses in the hippocampal cortex: Electron and light microscopic studies on the layer of mossy fibres. J. comp. Neurol. *117* (1961), 133—159.
3. Cajal, S., Ramon, Y., Histologie du système nerveux de l'homme et des vertèbres p. 993. Paris: Maloine. 1911.
4. Das, G. D., Hallas, B. H., Das, K. D., Transplantation of brain tissue in the brain of rat. I. Growth characteristics of neocortical transplants from embryos of different ages. Amer. J. Anat. *158* (1980), 135—146.
5. Gaarskjaer, F. B., Danscher, G., West, M. J., Hippocampal mossy fibres in the regio superior of the European hedgehog. Brain Res. *237* (1982), 79—90.
6. Haug, F.-M. S., Heavy metals in the brain. A light microscopic study of the rat with Timm's sulphide silver method. Methodological considerations and cytological and regional staining patterns. Adv. Anat. Embryol. Cell Biol. *47* (1973), 1—71.
7. Haug, F.-M. S., Light microscopical mapping of the hippocampal region, the pyriform cortex and the corticomedial amygdaloid nuclei of the rat with Timm's sulphide silver method. I. Area dentata, hippocampus and subiculum. Z. Anat. Entwickl.-Gesch. *145* (1974), 1—27.
8. Kromer, L. F., Bjorklund, A., Stenevi, U., Intracephalic implants, a technique for studying neuronal interactions. Science *204* (1979), 1117—1119.
9. Laurberg, S., Zimmer, J., Aberrant hippocampal mossy fibres in cats. Brain Res. *188* (1980), 555—559.

10. Lund, R. D., Harvey, A. R., Transplantation of tectal tissue in rats. I. Organization of transplants and pattern of distribution of host afferents within them. J. comp. Neurol. *201* (1981), 191—209.
11. Oblinger, M. M., Das, G. D., Connectivity of neural transplants in adult rats, analysis of afferents and efferents of neocortical transplants in the cerebellar hemisphere. Brain Res. *249* (1982), 31—49.
12. Raisman, G., Ebner, F. F., Mossy fibre projections into and out of hippocampal transplants. Neuroscience 1983 (in press).
13. Raisman, G., Ebner, F. F., Hippocampal transplants demonstrate the ability of the adult brain to receive and produce mossy fibre connections. In: Achievements in restorative neurology (Dimitrijevic, M. R., Eccles, J. C., eds.). In press (1983).
14. Schlessinger, A. R., Cowan, W. M., Gottlieb, D. I., An autoradiographic study of the time of origin and the pattern of granule cell migration in the dentate gyrus of the rat. J. comp. Neurol. *159* (1975), 149—176.
15. Schlessinger, A. R., Cowan, W. M., Swanson, L. W., The time of origin of neurons in Ammon's horn and the associated retrohippocampal fields. Anat. Embryol. *154* (1978), 153—173.
16. Stenevi, U., Bjorklund, A., Svengaard, N.-A., Transplantation of central and peripheral monamine neurons to the adult rat brain: Techniques and conditions for survival. Brain Res. *114* (1976), 1—20.
17. Sunde, N., Zimmer, J., Transplantation of central nervous tissue. An introduction with results and implications. Acta neurol. Scand. *63* (1981), 323—335.

Author's address: Dr. G. Raisman, Laboratory of Neurobiology, National Institute for Medical Research, Mill Hill, London NW71AA, U.K.

Acta Neurochirurgica, Suppl. 32, 61—64 (1983)
© by Springer-Verlag 1983

*Departments of Neurology, ** Neuropathology, and *** Neuroradiology,
University of Hamburg, Federal Republic of Germany

Comprehensive Monitoring and Computerized Tomographic Follow Up in Patients with Acute Severe Head Injury: Coma-Outcome Correlations

By

R. W. C. Janzen*, W. Rohr*, P. Neunzig*, H. J. Colmant**,
and D. Kühne***

Summary

In 88 patients with acute severe head injury the Glasgow coma scale (GCS), computerized tomography and autonomic monitoring during the first 24 hours were used to characterize different types of post-traumatic brain dysfunction. A subgroup is defined—GCS 8.1—with signs of diffuse brain swelling and autonomic instability which included two cases with dorsal midbrain lesions. These preliminary results suggest that autonomic instability is a valid criterion for identifying patients suffering from diffuse axonal injury.

Keywords: Coma; head injury; autonomic dysfunction; diffuse axonal injury.

Introduction

Post-traumatic brain dysfunction depends on the type of primary brain damage[6]. Variations in post-traumatic coma can be described by monitoring the overall function of the CNS using the Glasgow coma scale[7]. For detailed clinico-pathologic correlations, however, additional criteria are required, *e.g.* the results of CT scan, monitoring of ICP and EEG[3,4]. The present study was focussed on the question to what extent patients suffering from the syndrome of diffuse axonal injury (DAI) can be identified. These patients are known to be characterized by persisting deep coma, bilateral extensor rigidity and autonomic dysfunction[1].

Material and Methods

A group of 100 consecutive patients in our neurologic intensive care unit with acute severe head injury (ASHI) was analyzed using the following criteria: On admission: 1. assessment of the Glasgow coma scale and the pupils; 2. a CT scan within the first 12 hours which was assessed for swelling (absent, unilateral, or diffuse) and the types of lesion present (none, focal lesions or lesions with mass effects); 3. monitoring of the autonomic state for the first 24 hours plotting every minute the mean value of the heart rate, respiration, blood pressure, and body temperature. Variations in the autonomic state over a one-day period were expressed by the variance coefficient after Pearson [8]. Treatment was graded from 1 to 5: Sedation; sedation + intubation; intubation + hyperventilation; intubation + hyperventilation + barbiturates; intubation + hyperventilation + barbiturates + osmotics.

Follow up included: 1. Glasgow outcome scale, 1 for dead to 5 for minimal disability [5], and 2. CT examination 2–3 weeks after injury classifying the brain: CSF ratio (normal, ventricular dilatation, or diffuse atrophy) and the outcome of the primary lesions (absent, small defects, a large defects) [4].

Complete data was available for 88 patients (29 females, 59 males; age 34.5 years [SD 20.7]). Homogeneity of grouping was tested by the correlation of coma: outcome.

Results

The coma-outcome correlation of the total group was not significant (Table 1 A), even when anisocoria was included as a focal sign of a midbrain lesion [2]. The next step, on the basis of the early CT scan was to exclude the eight patients with mass lesions. Brain swelling as a possible sign of DAI when followed by ventricular dilatation or diffuse atrophy was found in 53 patients (group II). This group differed from the reference group I (n = 27), and included the only 2 patients with focal lesions that were assumed to be located in the dorsal midbrain (Table 1 B). The next step was to measure autonomic function and calculate the variance coefficients after Pearson [8] in the cases in Group II. In 41 cases complete monitoring of autonomic parameters during the first 24 hours following the injury was available. From these, variation in the heart rate was selected as the most representative parameter. Its mean value was 11.4 (SD 3.6), showing no significant correlation with the coma scale in patients in group II. Comparing group II a (hr—vc below 11.4; Table 1 C) and group II b (hr—vc above 11.4) the coma-outcome correlation coefficient was significantly different (Table 1 C). Subgroup II b was characterized by GCS 8.1, diffuse brain swelling in the early CT scan and autonomic instability as shown by a high variance coefficient of the heart rate.

Table 1 A. *Coma-Outcome Correlation of the Total Group (n = 88)*

Coma scale	3–5	6–8	9–11	12–14	15	(%)
Outcome dead	2	6	3	5	1	19.3
vegetative	4	2				6.8
severely disabled	1	4	1	3	1	11.3
moderately disabled		9		5	2	18.1
non/minimal disability	3	12	11	9	4	44.3
(%)	11.5	37.5	17	25	9	

Table 1 B. *Excluding 8 Patients Suffering from Lesions with Mass Effects, Two Groups Were Differentiated According to the Early CT: Group I (n = 27) with no or Lateralized Swelling and Group II (n = 53) with Diffuse Brain Swelling; Coma-Outcome Correlation Group I: Group II. r = 0.42, p < 5%*

	I Mean	SD	II Mean	SD
Treatment	2.3	1.3	3.7	1.1
Age	34.8	22.5	25.8	13.9
Coma	9.7	3.6	8.1	3.0
Outcome	3.6	1.5	4.0	1.4

Table 1 C. *Subgrouping of Group II According to the Heart Rate-Variance Coefficient (hr-vc): Subgroup II a with hr-vc Below 11.4 and Subgroup II b with hr-vc Above 11.4; Coma-Outcome Correlation Subgroup II a: Subgroup II b. r = 0.5, p < 5%*

	II a Mean	SD	II b Mean	SD
Treatment	3.5	1.4	3.6	1.4
Age	33.5	17.1	31.0	16.3
Coma	9.2	4.0	8.1	3.6
Outcome	3.9	1.7	4.0	1.3
Heart rate var. coef.	8.1	1.7	13.2	3.6

Discussion

In this study an attempt has been made to use comprehensive monitoring of the autonomic functions as an additional criterion for subgrouping patients with ASHI associated with diffuse brain swelling. Autonomic instability is selective in distinguishing a subgroup II b that is homogeneous according to the coma-outcome correlation. Within subgroup II b two cases were found presenting with a focal midbrain lesion. We therefore conclude that autonomic dysfunction can be used as a valid sign of DAI[6]. DAI seems to be only one part within the spectrum of brain damage that leads to severe contusions of the dorsal brain stem [1]. From a clinical point of view DAI has to be regarded as a special type of primary brain damage leading to different brain lesions that are less severe and not correlated with an outcome scaled 1 or 2[1].

A prospective study is required to establish if further techniques of evaluation and monitoring of the autonomic functions may be helpful in identifying different groups of patients suffering from ASHI with different types of primary brain damage.

References

1. Adams, J. H., Mitchell, D. E., Graham, D. I., Doyle, D., Diffuse brain damage of immediate impact type. Its relationship to "primary brain-stem damage" in head injury. Brain *100* (1977), 489—502.
2. Berge, J. H. van den, Schouten, H. G. A., Boomstra, S., Little Grunen, S. van, Braakmann, R., Inter-observer agreement in the assessment of ocular signs in coma. J. Neurol. Neurosurg. Psychiat. *42* (1979), 1163—1168.
3. Bowers, S. A., Marshall, L. F., Outcome in 200 consecutive cases of severe head injury in San Diego county: A prospective analysis. Neurosurgery *6* (1980), 237—242.
4. Clifton, G. L., Grossman, R. G., Makela, M. E., Neurological course and correlated computerized tomography findings after severe closed head injury. J. Neurosurg. *52* (1980), 611—624.
5. Jennett, B., Bond, M., Assessment of outcome after severe brain damage. A practical scale. Lancet *1* (1975), 480—484.
6. Jennett, B., Teasdale, G., Management of head injuries. Philadelphia: F. A. Davis. 1981.
7. Teasdale, G., Jennett, B., Assessment of coma and impaired consciousness. A practical scale. Lancet *2* (1974), 81—83.
8. Claus, G., Ebner, H., Grundlagen der Statistik, p. 89. Frankfurt/Main-Zürich: Harri Deutsch. 1971.

Author's address: Prof. Dr. R. W. C. Janzen, Department of Neurology, University of Hamburg, Martinistrasse 52, D-2000 Hamburg, Federal Republic of Germany.

Acta Neurochirurgica, Suppl. 32, 65—67 (1983)
© by Springer-Verlag 1983

Institute of Neurological Sciences, Glasgow, Scotland, U.K.

The Neuropathology of the Vegetative State and Severe Disability After Non-Missile Head Injury

By

D. I. Graham, D. McLellan, J. H. Adams, D. Doyle, A. Kerr, and L. S. Murray

Summary

A full neuropathological examination was undertaken in 35 cases of head injury who survived at least one month and who were either vegetative or severely disabled. Diffuse axonal injury was found in 21 cases, extensive hypoxic damage in the neocortex in 16 and secondary damage to the brain stem in 10.

Keywords: Head injury; disability; vegetative state.

Introduction

The aim of this study was to determine the nature of brain damage in a group of patients who were either vegetative[6] or severely disabled[5] as a result of a non-missile head injury and who had survived at least one month.

Material and Methods

During the 15-year period 1965–1980, post-mortem examinations were undertaken on 35 cases who had been assessed by the Department of Neurosurgery as being vegetative (28 cases) or severely disabled (7 cases) as a result of a non-missile head injury. There were 29 males and 6 females, the age range was from 14 months to 71 years (average 35 years), and survival ranged from 1 month to 14 years (29 cases died within 12 months of injury). 16 cases had sustained a road traffic accident, 13 a fall, 5 an assault, and in one the cause of injury was not known. 26 cases did not have a lucid interval[8], 6 had had a partial lucid interval[8] and in 3 cases adequate information was not available.

A full neuropathological examination was undertaken in each case[2].

Results

There was evidence of diffuse axonal injury[1] in 21 cases. Diffuse axonal injury known also as diffuse degeneration of the cerebral white matter[9], shearing injury[7, 10], diffuse damage to white matter of immediate impact type[3] and diffuse white matter shearing injury[11], is characterized pathologically by focal lesions in the corpus callosum and in the dorsolateral quadrant or quadrants of the rostral brain stem and microscopic evidence of diffuse damage to axons.

Extensive hypoxic necrosis of the neocortex of the type described by Graham *et al.*[4] was present in 16 cases; in 9 cases it had an arterial distribution; in 3 cases it was most severe along the arterial boundary zones of the cerebral hemispheres; in 3 cases it was severe and diffuse of the type associated with an episode of cardiac arrest or status epilepticus; and in one case there was a mixed pattern combining arterial territory and boundary zone hypoxic damage.

In 18 cases there was evidence of previous tentorial herniation[2] and in 10 of these there was secondary damage in the brain stem.

There was overlap in the pathology. Thus of the 21 cases with diffuse axonal injury there were 3 in which there was additional hypoxic brain damage and 2 in which there was secondary damage to the brain stem; of the 16 cases with hypoxic brain damage there were 8 with secondary damage to the brain stem and 3 with diffuse axonal injury; and of the 10 cases with secondary damage to the brain stem there were 8 with additional hypoxic brain damage and 2 with diffuse axonal injury. There were therefore no cases where severe disability or the vegetative state could be attributed solely to damage to the brain stem secondary to a high intracranial pressure.

One case did not fall into any of the 3 main categories of brain damage: this patient had sustained extensive contusional injury complicated by intracranial sepsis and marked hydrocephalus.

Additional features in the 35 cases included a fracture of the skull in 17, cerebral contusions in all cases, and intracranial haematoma in 22.

Discussion

In this study diffuse axonal injury was the most frequent structural abnormality in patients who survived for at least one month vegetative or severely disabled after a non-missile head injury. This was consistent with the observation that none of the patients with diffuse axonal injury experienced a lucid interval suggesting that overwhelming damage to the brain was sustained at the time of injury. Hypoxic brain damage with or without secondary brain stem damage occurred in the absence of diffuse axonal injury indicating that these processes may also cause the vegetative state or severe disability after head injury. There is no evidence in this study to suggest that secondary damage to the brain stem in the absence of any other brain damage is a cause of severe disability or the vegetative state after head injury.

References

1. Adams, J. H., Graham, D. I., Murray, L. S., Scott, G., Diffuse axonal injury due to non-missile head injury in humans: an analysis of 45 cases. Ann. Neurol. *12* (1981), 557—563.
2. Adams, J. H., Graham, D. I., Scott, G., Parker, L. S., Doyle, D., Brain damage in fatal non-missile head injury. J. Clin. Pathol. *33* (1980), 1132—1145.
3. Adams, J. H., Mitchell, D. E., Graham, D. I., Doyle, D., Diffuse brain damage of immediate impact type: its relationship to "primary brain stem damage" in head injury. Brain *100* (1977), 489—502.
4. Graham, D. I., Adams, J. H., Doyle, D., Ischaemic brain damage in fatal non-missile head injuries. J. Neurol. Sci. *39* (1978), 213—234.
5. Jennett, B., Bond, M., Assessment of outcome after severe brain damage. Lancet *i* (1975), 480.
6. Jennett, B., Plum, F., Persistent vegetative state after brain damage. Lancet *i* (1972), 734—737.
7. Peerless, S. J., Rewcastle, N. B., Shear injuries of the brain. Can. Med. Assoc. J. *96* (1967), 577—582.
8. Reilly, P. L., Graham, D. I., Adams, J. H., Jennett, B., Patients with head injury who talk and die. Lancet *i* (1975), 375—377.
9. Strich, S. J., Diffuse degeneration of the cerebral white matter in severe dementia following head injury. J. Neurol. Neurosurg. Psychiat. *19* (1956), 163—185.
10. Strich, S. J., Shearing of nerve fibres as a cause of brain damage due to head injury. Lancet *ii* (1961), 443—448.
11. Zimmerman, R. A., Bilianiuk, L. T., Gennarelli, T. A., Computerized tomography of shearing injuries of the cerebral white matter. Radiology *127* (1978), 393—396.

Authors' addresses: D. I. Graham, Ph.D., F.R.C.Path., D. McLellan, M.B. Ch.B., J. H. Adams, Ph.D., F.R.C.Path., D. Doyle, M.D., A. Kerr, M.Sc., University Department of Neuropathology; L. S. Murray, M.Sc., University Department of Neurosurgery, Institute of Neurological Sciences, Southern General Hospital, Glasgow G51 4TF, U.K.

Acta Neurochirurgica, Suppl. 32, 69—73 (1983)
© by Springer-Verlag 1983

Institute of Forensic Medicine, Northwestern University School of Medicine and
Cook Conty, Chicago, Illinois, U.S.A.

Ponto-Medullary Avulsion Associated with Cervical Hyperextension

By

J. E. Leestma, M. B. Kalelkar, and S. Teas

With 2 Figures

Summary

19 cases of partial or complete ponto-medullary avulsion are reported. This
type of damage seems to be produced by severe hyperextension of the head on the
neck with or without an additional rotational component.

Keywords: Head injury; ponto-medullary avulsion.

Traumatic avulsion or laceration of the ponto-medullary (PM)
junction is regularly observed in forensic autopsy cases but is
thought by many pathologists to be artifactual, having been
produced by forceful removal of the brain[5]. Nevertheless this type
of lesion has been reported as a direct consequence of head and
neck trauma many times[2, 3, 5 − 8] and is the subject of another
presentation in this symposium by Pilz (see p. 75). We present an
analysis of 19 cases in which partial or complete separation of the
pons from the medulla occurred in association with a traumatic
death usually from some form of vehicular or railway accident.

These 19 cases were all "medical examiner's" cases and ranged in
age from 9 to 62 years. 18 were males and one was a female. 16 cases
were involved in vehicular or railway accidents. The three

remaining cases died as a result of being struck by a hydraulic lift, a homocidal beating and a suspected homocidal beating respectively. Eight cases had blood ethanol levels between 72 and 426 mg%. Other brain/head injuries included cervical/medullary separation (4 cases), avulsion of the septum pellucidum with intraventricular hemorrhage (9 cases), a high cervical fracture-dislocation (13

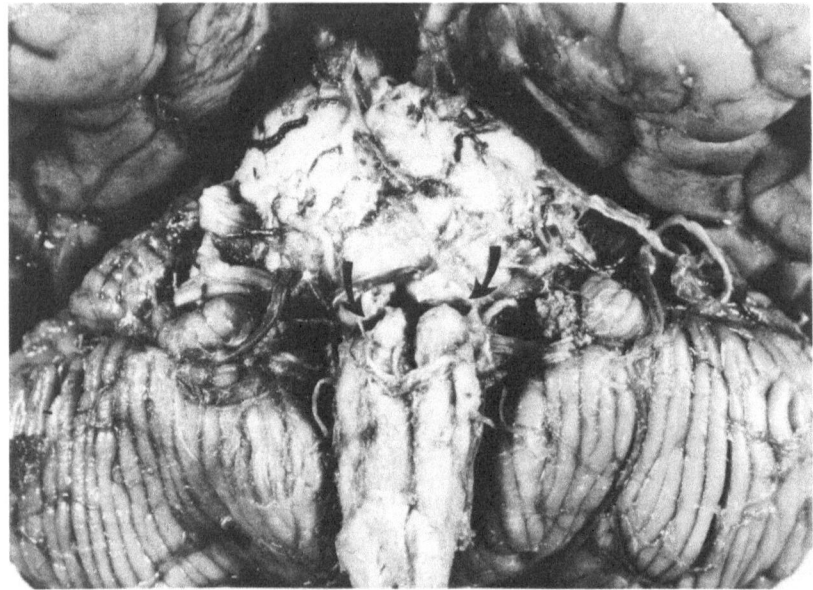

Fig. 1. A large nearly complete ponto-medullary avulsion is illustrated. Note the lack of significant hemorrhage in the region of the rent, a typical feature of this lesion

cases), a "ring" or other basal skull fracture (7 cases), and diffuse subarachnoid hemorrhage (9 cases). All but two cases appeared to have died very soon after their injury but one survived for 35 minutes and another lived for 90 minutes after injury. The victims had apparently been struck in about equal numbers from the rear, the front, and the side which probably produced at least some antero-posterior and most likely a violent rotatory movement of the head on the neck.

A typical lesion is illustrated in Fig. 1 where there is an almost complete separation of the medulla from the pons. A midline section of another PM avulsion which was incomplete is illustrated in Fig. 2. This photograph shows focal hemorrhage in the rent. Microscopic examination of the midline of the PM junction invariably demonstrates evidence of perivascular bleeding and occasional oedema in the torn areas, confirming that these lesions are ante-mortem and not artifactual.

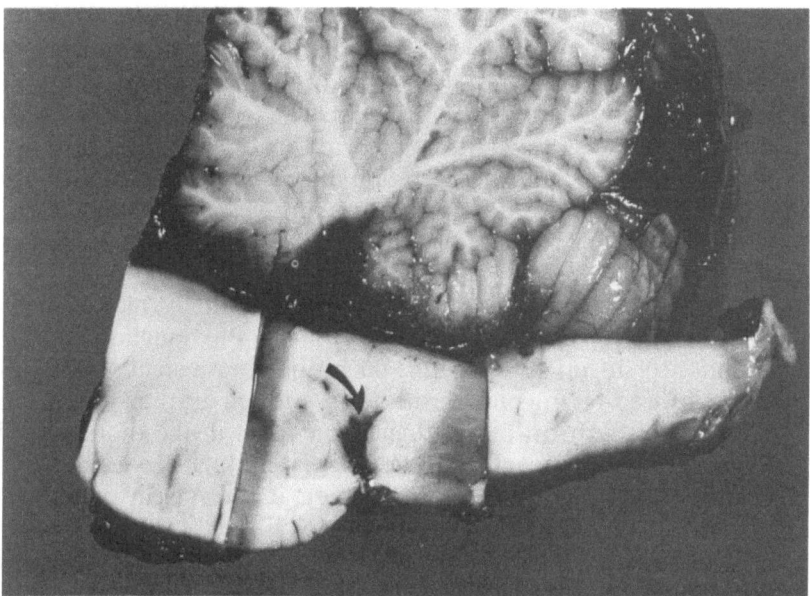

Fig. 2. A different case of PM avulsion reveals the focal hemorrhage which can usually be demonstrated if a median or paramedian section of the brain stem is made. Note also the subarachnoid and intraventricular haemorrhage which was caused by a co-existent avulsion in the fornix/septum pellucidum, not shown

The mechanism most likely responsible for this lesion is severe hyperextension of the head on the neck alone or coupled with a rotational component[5, 6, 8] and this may also result in fracture-dislocation of the upper cervical spine resulting in damage to the cervico-medullary junction as well. This concept is supported by the careful experimental work of Gennarelli and co-workers[1] who observed PM avulsion associated with sudden death in monkeys when their heads were rapidly accelerated in a special apparatus along the sagittal plane. We have studied an unusual case, not included in this study, which also supports the notion that cervical hyperextension may be mechanically important. This case involved a violent psychotic patient who was restrained from behind by an overzealous attendant by means of an "arm lock" (around the neck) which resulted in sudden paralysis and coma in the patient. Autopsy examination revealed a partial PM avulsion with little bleeding and no evidence of asphyxia. Clearly in this case hyperextension and possibly some rotational component were involved.

What may seem to be an unusual lack of massive hemorrhage in association with PM avulsion may be due to the fact that the PM junction lies between two vascular territories, the basilar artery and its branches (pons) and the vertebral artery and its branches (medulla) where a "watershed" region may exist[4]. Furthermore, no major avulsion or rupture of neighbouring vessels can be demonstrated on careful examination, though occasionally this can be the case[3].

PM avulsion usually results in rapid death by mechanisms probably related not just to tearing of the PM junction but to disruptive torsional forces acting on higher brain-stem structures which may lead to unconsciousness and sudden death in their own right, not to mention other traumatic events in the thorax, and abdomen which are usually severe. Coexistent high cervical lesions may impair respiration as well. However, it seems clear that some individuals who have their major injuries limited to the neck and lower brain stem may survive for extended periods[2,7] with PM lesions but these cases are not common.

Acknowledgements

We thank Dr. Robert J. Stein, the Medical Examiner of Cook County, and his staff at the Cook County Institute of Forensic Medicine, Chicago, Illinois, for their assistance in the collection of these cases.

References

1. Adams, J. H., Gennarelli, T. A., Graham, D. I., Brain damage in non-missile head injury: observations in man and subhuman primates. In: Recent advances in neuropathology, Vol. 2 (Smith, W. T., Cavanagh, J. B., eds.), pp. 167—190. Edinburgh-London-Melbourne-New York: Churchill-Livingstone. 1982.
2. Hardman, J. M., The pathology of traumatic brain injuries. In: Advances in neurology, Vol. 22 (Thompson, R. A., Green, J. R., eds.), pp. 15—50. New York: Raven Press. 1979.
3. Krauland, W., Verletzungen der intrakraniellen Schlagadern, pp. 20—37, 85—109. Berlin-Heidelberg-New York: Springer. 1982.

4. Leestma, J. E., Noronha, A., Pure motor hemiplegia, medullary pyramid lesion, and olivary hypertrophy. J. Neurol., Neurosurg., Psychiat. *39* (1976), 877—884.
5. Lindenberg, R., Freytag, E., Brainstem lesions characteristic of traumatic hyperextension of the head. Arch. Path. (Chicago) *90* (1970), 509—515.
6. Patscheider, H., Zur Entstehung von Ringbrüchen des Schädelgrundes. Dtsch. Z. ges. gerichtl. Med. *52* (1961), 243—281.
7. Pilz, P., Strohecker, J., Grobovschek, M., Survival after traumatic ponto-medullary tear. J. Neurol., Neurosurg., Psychiat. *45* (1982), 422—427.
8. Würmeling, H. B., Struck, F., Hirnstammrisse bei Verkehrsunfällen. Beitr. gerichtl. Med. *23* (1965), 297—302.

Authors' addresses: J. E. Leestma, M.D., Department of Pathology (Neuropathology), Northwestern University School of Medicine, Children's Memorial Hospital, 2300 Children's Plaza, Chicago, IL 60614, U.S.A.; M. B. Kalelkar, M.D., Office of the Medical Examiner, Cook County Institute of Forensic Medicine, 1828 West Polk Street, Chicago, IL 60612, U.S.A.; S. Teas, M.D., Office of the Medical Examiner, Cook County Institute of Forensic Medicine, 1828 West Polk Street, Chicago, IL 60612, U.S.A.

Acta Neurochirurgica, Suppl. 32, 75—78 (1983)

Neurological Institute, University of Vienna, Austria

Survival After Ponto-Medullary Junction Trauma

By

P. Pilz

With 2 Figures

Summary

Four cases with partial avulsions of the ponto-medullary junction resulting from a head injury with survival from 8 to 26 days are reported. Such lesions are not necessarily immediately fatal and they must be looked for specifically post-mortem.

Keywords: Head injury; primary brain-stem damage.

Introduction

Traumatic lesions of the ponto-medullary junction (PMJ) were first noticed by forensic pathologists[3, 4, 6] and they are now widely accepted as primary traumatic lesions in the central nervous system[2]. Nearly all patients so far reported died immediatly after the accident. The present communication describes 4 cases who survived this injury up to 26 days (two of the cases have been reported previously[5]).

Case Reports

All patients were female aged 10, 12, 43, and 67 years. After road traffic accidents they survived 8, 10, 17, and 26 days respectively. Post-traumatic coma lasted between 30 minutes and 10 hours, only one patient remaining comatose until death. All patients displayed paralysis of the abducens nerve on one side, and in two cases there was in addition unilateral paresis of the tongue and soft palate. Pyramidal tract lesions either severe or transient were present in all. In three cases there was subluxation of the craniocervical junction. Bruises to the forehead, chin or occiput were present in all patients but in no case there was a fracture of the skull. Death was due to extracranial causes in all.

Neuropathology

Neuropathological examination showed a deep tear of the ventral PMJ in one case which approached the floor of the 4th ventricle (Fig. 1). In another case there

Fig. 1. Deep tear at the ponto-medullary junction surrounded by a broad necrotic zone (Klüver Barrera)

was only a small rent a few mm deep while in the remaining two cases there was a wedge of necrosis at the PMJ (Fig. 2). These small lesions could only be identified in sagittal sctions. In addition small focal infarcts and haemorrhages were observed in the distribution of the perforating arteries in the pons and medulla. Ruptured small arteries and veins were found in the meninges over the pyramids and accounted for subarachnoid haemorrhage in this region. A striking feature was the presence of numerous axonal retraction balls not only adjacent to the tear but disseminated throughout longitudinal fibres of the entire brain stem as far as the cerebral peduncles. Wallerian degeneration occured in the clinically affected cranial nerve roots, In no case were cortical contusions found. Severe anoxic brain damage occured in one case who was resuscitated after the accident and remained comatose until death.

Discussion

Traumatic tears at the PMJ must be separated from artefactual tears which frequently occur in this region post mortem[3, 6]. Cranial nerve lesions result from stretching of the nerve roots, the 6th nerve

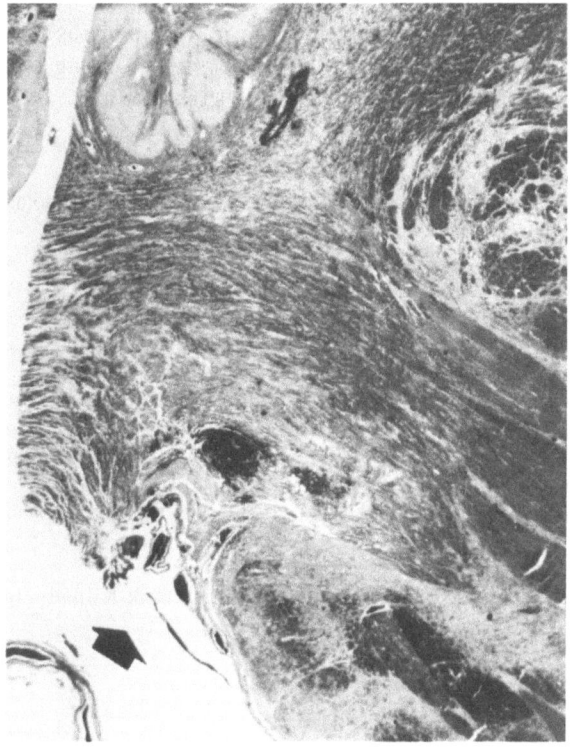

Fig. 2. Necrotic zone and small focal haemorrhages at the ponto-medullary junction (Klüver Barrera)

frequently being severed at the necrotic zone at the PMJ. Pyramidal tract lesions are due not only to the tear or necrosis but also to wide-spread axonal injury.

The neuropathological changes result from downward displacement of the brain stem at the time of injury and occur as a result of axial traction of the cervical spine on the head[4, 6] or acute hyperextension of the head[2, 3]. The latter is the most frequent cause of tears at the PMJ. Severe axonal injury in the brain stem is also due to axial traction, which might be the cause of post-traumatic unconsciousness [1].

The cases reported clearly demonstrate that tears at the PMJ are not always immediately fatal. The clinical syndrome consists of unilateral abducens nerve palsy, a pyramidal tract lesion, and injuries of the cranio-cervical junction along with a severe concussion syndrome.

Neuropathological changes (a tear or necrosis at the PMJ, axonal injury, small focal infarcts, and haemorrhages in the pons and medulla) must be sought in sagittal sections.

References

1. Friede, R. L., Experimental concussion accleration. Pathology and mechanics. Arch. Neurol. (Chic.) 4 (1961), 449—462.
2. Hardmann, J. M., Pathology of traumatic brain injuries. In: Advances in neurology, Vol. 22 (Thompson, R. H., Green, J. R., eds.). New York: Raven Press. 1979.
3. Lindenberg, R., Freytag, E., Brain stem lesions characteristic of traumatic hyperextension of the head. Arch. Pathol. 90 (1970), 509—515.
4. Patscheider, H., Zur Entstehung von Ringbrüchen des Schädelgrundes. Dtsch. Z. ges. gerichtl. Med. 52 (1961), 13—21.
5. Pilz, P., Strohecker, J., Grohovschek, M., Survival after traumatic ponto-medullary tear. J. Neurol., Neurosurg., Psychiat. 45 (1982), 422—427.
6. Würmeling, H. B., Struck, F., Hirnstammrisse bei Verkehrsunfällen. Beitr. gerichtl. Med. 23 (1965), 297—302.

Author's address: Dr. P. Pilz, Landes-Nervenklinik Salzburg, Ignaz-Harrer-Strasse 79, A-5020 Salzburg, Austria.

Acta Neurochirurgica, Suppl. 32, 79—85 (1983)

Department of Neuropathology University of Zagreb Clinical Medical Centre,
Zagreb, Yugoslavia

Traumatic Tears of the Tela chorioidea: A Hitherto Unrecognized Cause of Post-Traumatic Hydrocephalus *

By

N. Grčević

With 2 Figures

Summary

A new and hitherto unrecognized phenomenon of rupture of the tela chorioidea in closed head injury of acceleration-deceleration type is described. It occurs very frequently, especially in association with blows in the centro-axial plane even if the acceleration forces are relatively mild. These tears are regularly followed by intraventricular bleeding which follows the flow of the CSF into the subarachnoid space producing a leptomeningeal reaction with impairment of absorption of CSF and consequent communicating hydrocephalus.

Keywords: Tears of choroid tela; traumatic rupture of the choroid tela; traumatic ventricular bleeding; post-traumatic hydrocephalus; normal pressure hydrocephalus.

Introduction

It is known that due to certain specific physical mechanisms in closed head injury of acceleration-deceleration type the "centro-axial" cerebral regions are especially vulnerable and suffer severe

---

* This investigation was supported, in part, by Research Grant No. 02-025-1, N. I. H., D. H. E. W. of the U.S. Government, Bethesda, Maryland, U.S.A.

damage[1, 8, 9, 12, 15, 17, 19, 20]. This is particularly true if traumatizing forces act along the longest diameter of the skull.
Certain periaxial structures are especially frequently involved and
the lesions there produce a peculiar pattern of "inner cerebral
trauma"[8, 9]. In the course of a systematic study on this pattern of
lesions[8], we observed a strikingly high incidence of tears in the tela
chorioidea of the 3rd and lateral ventricles with consequent
haemorrhage into the plexus and into the ventricular system. Since
lesions of this type have not been previously described in cranio-
cerebral trauma, we undertook further investigations resulting in
observations which we are presenting here as a preliminary
communication.

Methods and Material

The material for this study consisted of 66 brains from individuals who
survived a closed head injury from a few minutes to 846 days. 78% of the cases
were due to traffic accidents; the rest were falls (15.2%) and blows by a blunt object
(6%). In the group of traffic accidents there was a predominance of traumatic
forces in an axial direction with 73% frontal and 17.6% occipital blows, making a
total of 90.6% with centroaxial blows. One case sustained a blow on the vertex
while only 4 cases resulting from traffic accidents (7.6%) suffered lateral blows.

All brains, removed at autopsy, were fixed in 10% formalin, embedded in
paraffin and examined by the method of subserial hemispheric sections stained by
the standard histological techniques used in neuropathology.

Results

As reported in detail elsewhere[8], our study showed a very high
incidence of lesions with the pattern of "inner cerebral trauma".
Within this pattern, lesions of the choroid tela in the 3rd and lateral
ventricles were an extremely frequent finding. The principal feature
of these lesions were tears and ruptures of the tela at the sites of its
insertion into the thalamus (Figs. 1 and 2) and ruptures of the veins
located at these sites (Figs. 1 and 2). These ruptures were followed
by bleeding into the reticular tissue of the tela and the choroid
plexuses (Figs. 2 a and b), and into the ventricular system. The
amount of haemorrhage varied from case to case and could not be
exactly defined because of the various times of survival, which
allowed the intraventricular blood to be already washed out in the
subacute cases. However, traces of blood were almost always found
by careful histological examination. In acute cases, tears in the tela
and haemorrhages could be easily seen by naked eye inspection. In

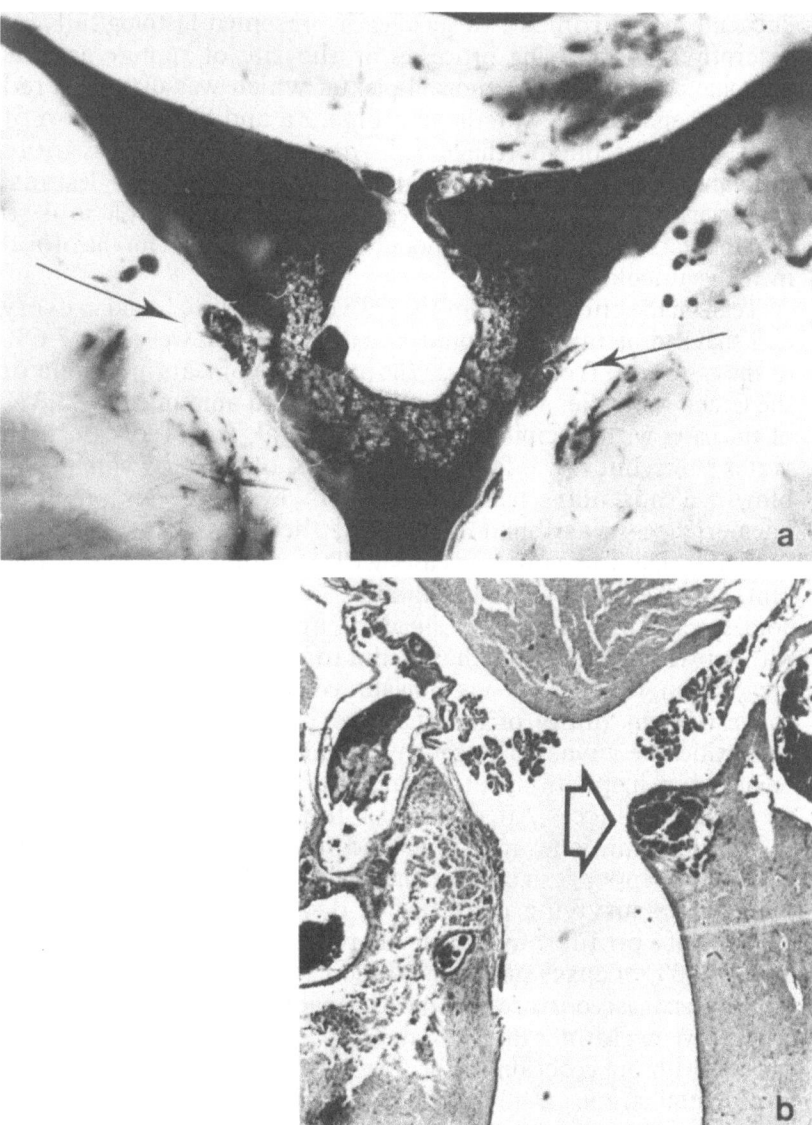

Fig. 1. *a* The tela of the 3rd ventricle has been torn from the stria medullaris thalami at the site of the vena cerebri interna, which also show traumatic damage bilaterally (arrows). The choroid plexus is detached, infiltrated with blood and floats freely. There is blood in the ventricles. 10 hours survival. *b* Histological section of the same case. Bleeding within the stria medullaris (arrow) due to tearing forces

subacute and chronic cases such tears presented histologically as resorptive and scarring processes at the site of rupture and as atrophic changes in the choroid plexus which was often severed from its connections with the tela (Figs. 2 a and b). Correlation of intraventricular haemorrhage with rupture of the tela was more difficult in the acute cases with other periventricular lesions. However, quite a few cases showed no evidence of other lesions so that the tears of the tela were undoubtedly responsible for the blood in the ventricular system.

Tears in the choroid tela of the 3rd ventricle were found in every case that had sustained occipital or vertex blows, as well as in 97.6% of the cases with frontal blows. The incidence of tears in the tela of the lateral ventricles was somewhat lower and amounted to 83.3% of the cases with occipital, 74% with frontal, and every case with vertex blows, but it was found in only 5.8% of the cases with lateral blows. Ventricular haemorrhage—or evidence of previous haemorrhage—was found in 84.1% of the cases.

In all acute and subacute cases with a rupture of the tela and ventricular bleeding there was always some blood in the subarachnoid space. This followed a standard and characteristic pattern with spread of blood from the cisterna magna symmetrically along the Sylvian fissures to accumulate over the parieto-temporal convexities. In many of the subacute cases with blood on the convexities there was no longer any blood in the ventricular system on naked eye inspection, but only a slight pink discolouration of the ependymal surfaces. After a further few days the blood on the convexities diminishes in amount and moves towards the sagittal region. It disappears in cases of longer survival. However, in 88.6% of the cases surviving more than 5 days there was histological evidence of a proliferative arachnoid reaction to the haemorrhage, and in 93.3% of cases surviving between 15 and 89 days there was a definitive adhesive process in the pia-arachnoid system. All of the cases surviving longer than 90 days (12 cases) had a severe adhesive process with obstruction to the flow of CSF. In several cases the parasagittal areas were particularly affected, the Pacchionian granulations being embedded in the proliferative process.

Hydrocephalus was present in every case surviving longer than 15 days (27 cases) as well as in 9 cases surviving between 5 and 14 days (53% of the total of 17 cases from this group). In spite of varying degrees of a diffuse sclerotic process of the white matter in several of the cases surviving more than 90 days, an *ex vacuo* mechanism could not be considered responsible for severe hydro-

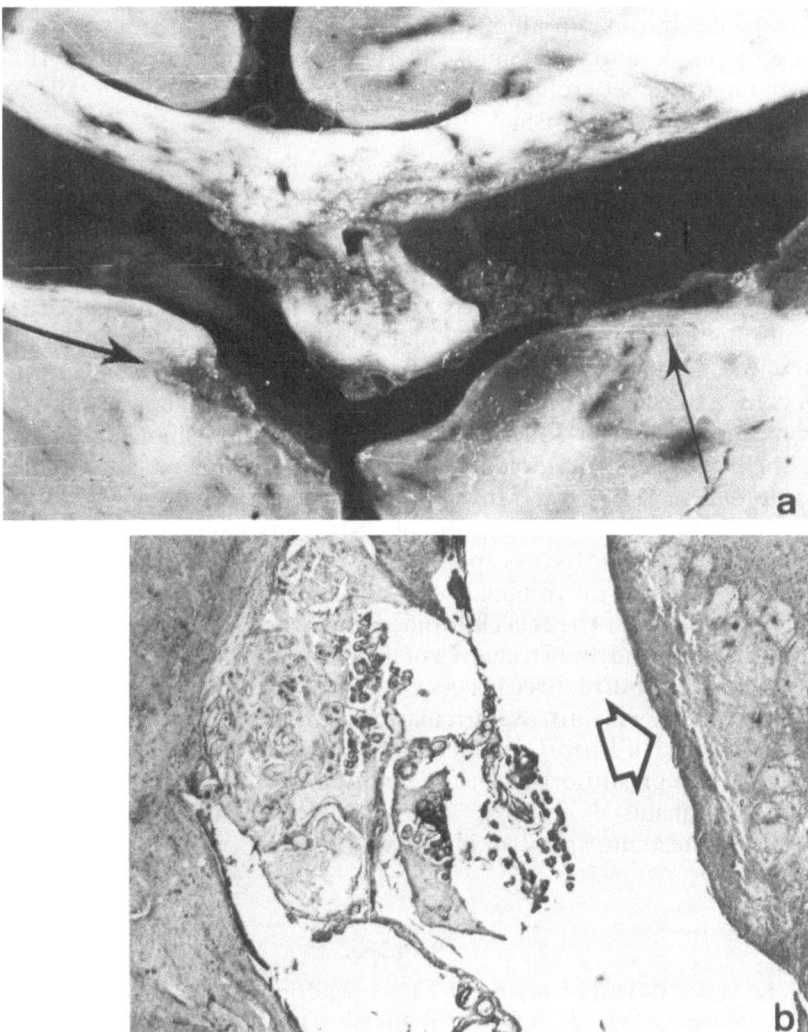

Fig. 2. *a* The tela chorioidea ventriculi lateralis is completely separated from its tenia on the left (thicker arrow) and there is rupture of the vein on the right side with haemorrhage underneath the tenia thalami and into the plexus. The attachment of the tela chorioidea ventriculi tertii to the stria medullaris is torn; the plexus is extremely retracted and reduced to a small granule at the lower surface of the fornix. The whole process is subacute and the sites of rupture are in the stage of resorption and scarring. Survival 23 days. *b* Histological section of the tela and plexus of the lateral ventricle of the same case. Note atrophy of the plexus and obvious necrosis at the thalamic site of the previous attachment from which the tela was torn (arrow)

cephalus because in all of such cases arachnoid adhesions were so pronounced and communicating hydrocephalus so obvious that pathogenetic correlations were quite clearly in favour of an obstructive mechanism.

Discussion

Our observations clearly indicate that there is a very high incidence of tears in the choroid tela in closed head injury of acceleration-deceleration type, especially if the traumatizing forces act along the longest diameter of the skull. This phenomenon is part of the pattern of "inner cerebral trauma" and may be biomechanically explained in the same manner [8, 9, 15, 20]. The ventricular haemorrhage that follows is followed by subarachnoid spread of the blood. Subarachnoid haemorrhage of any origin may trigger a leptomeningeal proliferative and adhesive process [4, 5, 11, 13, 17, 22], which, by blocking absorption of CSF, may cause an obstructive communicating hydrocephalus [2, 3, 7, 14, 21, 22]. Our experience suggests that even minimal blows in a centro-axial direction may produce tears in the tela chorioidea since they are the most delicate of all of the midline structures of the brain. The sequence of events after such ruptures, regardless of their severity always follows the same course viz. intraventricular bleeding followed by subarachnoid spread of blood and an adhesive arachnoidal reaction. Thus, one could presume that in the pathogenesis of "normal pressure hydrocephalus" [2, 3, 6, 10, 16, 18], such "occult" ruptures of the choroid tela may play an important role.

References

1. Adams, J. H., The neuropathology of head injuries. In: Hdbk. of clinical neurology, Vol. 23 (Vinken and Bruyn, eds.), pp. 35—65. Amsterdam: North-Holland. 1975.
2. Adams, R. D., Recent observations on normal pressure hydrocephalus. Schweiz. Arch. Neurol. Neurochir. Psychiat. *116* (1974), 7—15.
3. Blaylock, R. L., Kempe, L. G., Hydrocephalus accociated with subarachnoid hemorrhage. Neurochirurgie *21* (1978), 20—28.
4. Doppner, Th., Spaar, F. W., Orthner, H., Zur Neuropathologie des posttraumatischen Hirndrucks im Kindesalter. Zugleich ein Beitrag zur Klinik und Pathogenese der „wachsenden Schädelfraktur". Z. Neurol. *202* (1972), 37—51.
5. Ellington, E., Margolis, G., Block of arachnoid villus by subarachnoid hemorrhage. J. Neurosurg. *30* (1969), 651—657.

6. Foroglou, G., Zander, E., Nos expériences avec l'hydrocéphalie interne communicante. Arch. Suisse Neurol. Neurochir. Psychiat. *115* (1974), 291—303.
7. Galera, R., Greitz, T., Hydrocephalus in the adult secondary to the rupture of intracranial arterial aneurysms. J. Neurosurg. *32* (1970), 634—641.
8. Grčević, N., Topography and pathogenic mechanisms of lesions in "inner cerebral trauma". Rad Yug. Acad. Sci. *402/*18 (1982), 265—331.
9. Grčević, N., Jacob, H., Some observations on pathology and correlative neuroanatomy of sequels of cerebral trauma. In: Proc. 8th Intern. Congr. Neurology, Vol. 1, pp. 369—373. Wien: 1965.
10. Hakim, S., Adams, R. D., The special clinical problem of symptomatic hydrocephalus with normal CSF pressure. J. neurol. Sci. *2* (1965), 307—327.
11. Iwanowski, L., Olszewski, J., The effects of subarachnoid injections of iron-containing substances on the nervous system. J. Neuropath. exp. Neurol. *19* (1960), 433—448.
12. Jellinger, K., Protrahierte Formen der posttraumatischen Encephalopathie. Beitr. gerichtl. Med. *23* (1965), 65—118.
13. Kessel, K., Traumatische Subarachnoidalblutung. In: Neurotraumatologic. Vol. I (Kessel, ed.), pp. 180—191. München-Berlin-Wien: Urban and Schwarzenberg. 1969.
14. Kibler, R. F., Couch, R. S. C., Crompton, M., Hydrocephalus in the adult following spontaneous subarachnoid hemorrhage. Brain *84* (1961), 45—60.
15. Lindenberg, R., Trauma of meninges and brain. In: Pathology of the nervous system, Vol. 2 (Minkler, ed.), pp. 1706—1765. New York: McGraw-Hill. 1971.
16. Ojeman, R. G., Normal pressure hydrocephalus. In: Scientific foundations of neurology (Critchley *et al.,* eds.), pp. 302—323. London: W. Heinemann Med. Books Ltd. 1972.
17. Peters, G., Pathologische Anatomie der Verletzungen des Gehirns und seiner Häute. In: Neurotraumatologie, Vol. 1 (Kessel, ed.), pp. 31—117. München-Berlin-Wien: Urban und Schwarzenberg. 1969.
18. Rudd, Th. G., O'Neal, J. T., Nelp, W. B., CSF circulation following subarachnoid hemorrhage. J. Nucl. Med. *12* (1971), 61—63.
19. Ule, G., Dohner, W., Bues, E., Ausgedehnte Hemisphärenmarksschädigung nach gedecktem Hirntrauma mit apallischem Syndrom und partieller Spätrehabilitation. Arch. Psychiat. *202* (1961), 155—176.
20. Unterharnscheid, F., Die gedeckten Schäden des Gehirns. In: Monogr. a. d. ges. Geb. der Neurol. u. Psychiat. Vol. 103 (Spatz *et al.,* eds.). Berlin-Göttingen-Heidelberg: Springer. 1963.
21. Wieser, H. G., Probst, Ch., Beitrag zur Klinik des Hydrocephalus mit besonderer Berücksichtigung des posttraumatischen Hydrocephalus male resorptivus und der Shuntindikation. J. Neurol. *212* (1976), 1—21.
22. Zander, E., Die posttraumatische Hydrozephalie und ihre Behandlungsmöglichkeiten. Schweiz. med. Wschr. *99* (1969), 1624—1629.

Author's address: Prof. N. Grčević, M.D., Department of Neuropathology, Clinical Medical Centre Rebro, Yu-41000 Zagreb, Yugoslavia.

Acta Neurochirurgica, Suppl. 32, 87—90 (1983)
© by Springer-Verlag 1983

Department of Pathology, University of Manchester, Manchester, U.K.

Birth Injury to the Cervical Spine and Spinal Cord

By

Helen Reid

With 1 Figure

Summary

Mechanical trauma to the cervical spine still occurs at birth. In 2 of 48 perinatal post-mortems traumatic damage to the cervical spinal cord was found. Also in this series at least 12% of cases from one hospital showed some degree of trauma to the cervical spine but this was of a lesser degree in individual cases than 20 years ago.

Keywords: Birth; spine; spinal cord; trauma.

Introduction

It is well known that the brain may be damaged by mechanical forces during birth, tentorial tears and subdural haemorrhage being found at autopsy. Changes in obstetrical management have caused this to decrease but not to disappear. Similarly mechanical force is known to damage the cervical cord and spine. Obstetricians however are also acutely conscious of the dangers of birth asphyxia and the need to expedite a birth if this occurs but, if too much force is applied, this can also lead to death through mechanical trauma.

Twenty years ago Yates[2] found that damage to the cervical spine was fairly common in a series of still-births and neonatal deaths, 28% having distortional trauma to the cord. In a series of 430 cases Towbin[1] in 1969 found 10% with spinal and brain stem injury. As obstetrical practices have improved and there are fewer full-term infants coming to autopsy compared to premature infants, I have again examined a series of cervical spines and brains from still-born and neonatal deaths to see if there are any changes in the incidence of spinal damage.

Methods and Material

There were 48 cases, 43 of which were from a maternity hospital where there were 83 autopsies on 101 deaths in one year. This hospital is a regional centre and premature infants are transferred to it. The still-birth and perinatal mortality for the hospital births during that year was 1.75%.

The cervical spine from C 7 to the foramen magnum was removed. After decalcification, horizontal segments were processed and examined. Several areas of the brain were examined histologically and neuronal ischaemic cell change assessed.

In this series there were 18 infants born near term and altogether 14 were delivered by caesarian section. It was also ascertained that some of the premature infants had had a precipitous birth.

Results

The main findings are summarized in the table. One of the cases with spinal cord damage was due to cord compression by an epidural haemorrhage; in the other case there was haemorrhage and necrosis in the dorsal horns (Fig. 1) and there was also gliosis in one cortico-spinal tract because of traumatic damage to the pons.

In four of the cases with damage to a vertebral artery, tentorial tears were also present indicating that mechanical trauma was the cause of death. Epidural and subarachnoid haemorrhage in the spinal canal were the indicators used for distortional damage to the spine.

In only two of the cases was there neuronal ischaemic cell change in the brain; one who had had a difficult forceps delivery showed these changes in neurons scattered throughout the cerebrum and brain stem; in the other case they were seen only in the neurons of the basal ganglia. Periventricular leukomalacia was found in four cases, two of whom were delivered by caesarian section for antepartum haemorrhage. The other two were breech deliveries.

Discussion

On comparing this series with that of Yates[3] (Table 1), there has been no decrease in the number of cases with trauma to the cervical spine, but there has been a decrease in the amount of damage done in each individual case. The survey was carried out in the same hospital in both series so they are comparable to some extent.

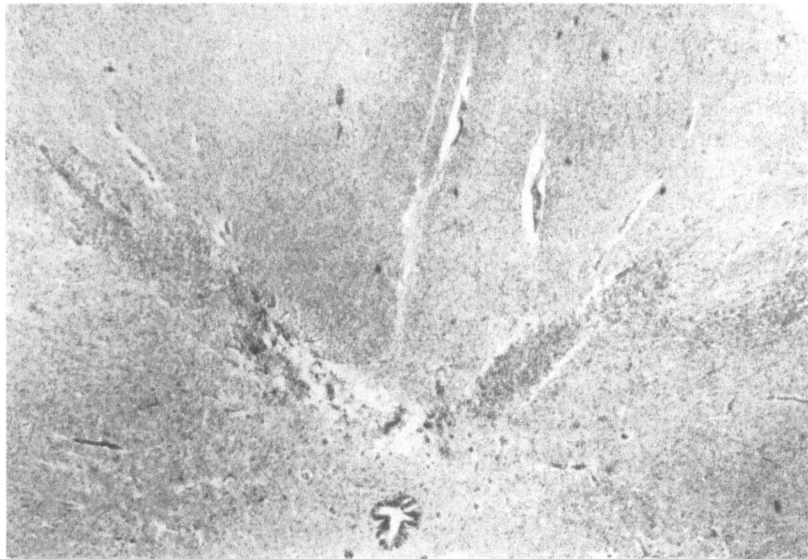

Fig. 1. The dorsal cervical spinal cord showing necrosis and haemorrhage in the dorsal horns. × 30

Table 1

Number of cases	This series 48	Yates[3] 250
Spinal cord damage	2 (4%)	4 (1.6%)
Tearing and bruising of cervical nerve roots	12 (25%)	25 (10%)
Vertebral artery damage	7 (14.5%)	42 (16.4%)
Distortional damage to the cervical spine	25 (52%)	72 (28.8%)

Yates[2] suggested that damage to a vertebral artery may be related to vascular problems in the brain stem causing various cranial nerve palsies and the atactic cerebellar form of cerebral palsy. This study does not support this hypothesis; on the contrary, I agree with Towbin's[1] view that direct mechanical trauma to the brain stem and cerebellum accounts for such neurological problems.

Mechanical trauma to the skull and spine undoubtedly still occurs and can cause death. In some of these infants other factors were also present such as congenital malformations. Also eight cases had a definite history of intrapartum asphyxia and only one was devoid of any cervical spine or cord damage. This underlines obstetricians' difficulties since death through intrapartum asphyxia can occur if the baby is not delivered quickly, and if too much force is used to deliver such an infant death as a result of mechanical trauma can occur, even when caesarian sections are carried out. It is also suggested that intrapartum asphyxia can predispose an infant to mechanical trauma.

In this series the number of infants with cord damage, torn cervical nerve roots and vertebral artery damage amounted to 12 of the 48 cases and these were all from one hospital and, therefore, at least 12% of all deaths from this hospital showed spinal cord or cervical spine damage in one year.

References

1. Towbin, A., Latent spinal cord and brain stem injury in newborn infants. Dev. Med. Child Neurol. *11* (1969), 54—68.
2. Yates, P. O., Birth trauma to the vertebral arteries. Arch. Dis. Child. *34* (1959), 436—441.
3. Yates, P. O., Perinatal injury to the neck and extracranial cerebral arteries. Proceedings 4th Internat. Cong. Neuropath. (Budapest) IV 1961, 20—23.

Author's address: Dr. Helen Reid, Lecturer in Neuropathology, Department of Pathology, Medical School, University of Manchester, Manchester M13 9PT, U.K.

Acta Neurochirurgica, Suppl. 32, 91—94 (1983)
© by Springer-Verlag 1983

Neuroradiology of the Sequelae of Spinal Cord Trauma

By

M. Perovitch

Summary

More than 850 patients with acute injuries to the spinal cord were explored by means of neuroradiological procedures (1954–1981). The sequelae of cord damage were assessed in 125 of these patients and the results of neuroradiological examinations correlated with the surgical or autopsy findings. Such opportunities to demonstrate the changes that follow the acute phase of a spinal cord injury are infrequent because of a lack of centralized team care and continuity of follow-up. Chronic post-traumatic changes, induced by mechanical kinetic or non-mechanical energies, were demonstrated after a long period of time, occasionally after two to ten years. They were classified into six groups. Disturbances of arterial circulation are emphasized as a major source of aggravation of the post-traumatic state of the spinal cord.

Keywords: Spinal cord; trauma; sequelae; neuroradiology.

In the period from 1954 to the end of 1981, we explored by means of neuroradiological procedures more than 850 patients with acute injuries to the spinal cord brought about by a blunt impact in traffic accidents, falls and sporting accidents, or by penetrating wounds. Patients whose spinal cord had been irradiated, exposed to the effect of some chemical compounds such as contrast media or spinal anaesthetics, or had sustained long-lasting compression are also included.

Out of these 850 patients we were able to examine 125 in the chronic phase of the spinal cord injury and to compare the neuroradiological findings with subsequent pathological or surgical exploration. Opportunities to demonstrate the changes which succeed spinal cord trauma are infrequent because cohesive team action and comprehensive care following the acute phase of the spinal cord injury are not as widely practised as they should be. This concept is now providing the grounds for meaningful co-ordinated research programmes aimed at a better understanding of the pathophysiology of spinal cord damage with the aim of retaining and improving all remaining neural functions.

Apart from radiography and tomography of the spine, the neuroradiological procedures used to establish the presence of post-traumatic alterations in the spinal cord included myelography with insoluble and soluble contrast media, gas myelography, spinal cord angiography, and, more recently, computed tomography and digital subtraction angiography (intravenous or intra-arterial). Since our experience with the application of nuclear magnetic resonance or positron emission tomography to the examination of the spinal cord is, at present, limited it is not encompassed in this study.

The chronic post-traumatic changes were classified into: (i) an intramedullary cavity (syrinx) with or without a foreign body; (ii) upward cavitation from the site of injury (post-traumatic syringomyelia) with occasionally downward expansion; (iii) atrophy of the spinal cord; (iv) adhesions; (v) vascular alterations; and (vi) the effects of a post-traumatic encapsulated extramedullary haematoma or cyst, tear of the dentate ligament, etc. Some of these sequelae were often combined.

About three weeks after trauma to the spinal cord, the acute changes subside and an intermediate period begins which is characterized by the disappearance of oedema and absorption of necrotic tissue and blood. Astrocytic gliosis leads to scar formation. Chronic post-traumatic changes take place after a longer period of time, occasionally after as long as two to ten years.

Longitudinal cavities within the spinal cord which replace the area of haemorrhagic necrosis in the acute stage of the injury could be demonstrated clearly in 19 cases particularly by means of gas-myelography or by injecting oxygen directly into the syrinx. Computed tomography was also used very successfully. The intramedullary cavity may contain a foreign body for a long period of time. We were able to demonstrate the presence of a bullet using positive or negative myelography in seven patients. By tilting the table up and down we were able to assess the extent of the free movements of the metallic foreign body in the cavity, and outline precisely its shape and size[1]. In 14 patients metrizamide and gas-myelography as well as computed tomography proved to be precise neuroradiological techniques for the evaluation of post-traumatic syringomyelia. With gas-myelography there was extensive filling of the cavities in the cord with oxygen, and a reduction of up to 85% of the remaining spinal cord tissue.

Angiographic studies performed one or more years after the acute phase of a spinal cord injury demonstrated vascular occlusions and aneurysmal formations or arteriovenous shunts similar to arteriovenous malformations. Spinal angiography was performed in 15 patients in the chronic post-traumatic phase. Impairment of the blood flow to an injured spinal cord may lead to progressive atrophy of its substance above and below the site of the original injury. This atrophy was well demonstrated by means of gas-myelography and computed tomography in 27 patients.

The formation of extensive scar tissue was seen particularly when there had been a penetrating wound. This type of injury is commonly accompanied by severe epidural, subdural, sub-arachnoid or spinal cord haemorrhage. Damage to the blood vessels of the spinal cord is usually severe, and hemisection or transection of the spinal cord may occur. In 12 patients thick scars encompassing long segments of the atrophic spinal cord were seen on gas-myelograms and computed tomograms. The spinal cord appears to enter a tube, and we call this appearance "the spinal cord in the boot".

Post-traumatic abscess formation—a rather rare phenomenon today, nerve root avulsion combined with a meningeal tear, cyst formation, and dislocations of fractured vertebrae as well as encapsulated or fibrous epidural haematoma, can affect the spinal cord and its blood supply for long periods of time leading to its atrophy. Similarly, herniated fragments of ruptured intervertebral discs, in particular in the cervical and thoracic regions, or post-traumatic osteophytes can produce chronic compression of the spinal cord clearly visible on computed tomograms or gas-myelograms (18 patients).

Following radiation therapy, for example for a lung carcinoma, with high doses of 4,000–6,000 rad or more, radiation myelitis may occur. In a previous study dealing with the problem of chronic radiation myelitis, we emphasized the pathological changes which may occur in the spinal cord, especially vascular occlusions, necrotic areas, and ultimately diffuse atrophy of the spinal cord. In eight patients these changes were demonstrated several years after radiation therapy[2].

The sequelae of injuries to the spinal cord by chemical

compounds have been seen after the subarachnoid injection of some types of contrast media or other medications. They were associated with extensive scar formation around the spinal cord[3]. The need for spinal cord injury centres has been recognized only recently. Through the formation of such centres it is becoming possible to use to full advantage accurate neuroradiological procedures that disclose during life the pathological processes which may evolve after an injury to the spinal cord.

References

1. Perovitch, M., Injuries to the spinal cord and meninges. In: Radiological evaluation of the spinal cord, Vol. 1 (Perovitch, M., ed.), pp. 143—169. Boca Raton, Florida: CRC Press. 1981.
2. Perovitch, M., Postradiation myelopathy. In: Radiological evaluation of the spinal cord, Vol. 2 (Perovitch, M., ed.), pp. 85—87. Boca Raton, Florida: CRC Press. 1981.
3. Perovitch, M., Complications related to positive contrast myelography. In: Radiological evaluation of the spinal cord, Vol. 1 (Perovitch, M., ed.), pp. 79—92. Boca Raton, Florida: CRC Press. 1981.

Author's address: Prof. M. Perovitch, M.D., The University of Connecticut School of Medicine, Farmington, CT 06032, U.S.A.

Acta Neurochirurgica, Suppl. 32, 95—97 (1983)
© by Springer-Verlag 1983

Institute of Neurological Sciences, Glasgow, Scotland, U.K.

A Computerized Data Retrieval System for Investigation of Brain Damage in Non-Missile Head Injury

By

Grace Scott, Audrey Kerr, and Lilian S. Murray

Summary

An analysis of brain damage in head injury using a computerized data retrieval system has contributed to the presentation of the various types of brain damage objectively, and has shed light on contre coup contusions and diffuse axonal injury.

Keywords: Head injury; quantitative analysis of brain damage.

Introduction

Brain damage in cases of non-missile head injury has in the past tended to be assessed and presented on a descriptive basis. This has led to difficulty in presenting objectively the findings in large series of cases and in comparing different subgroups of a series. In an attempt to achieve this a method has been devised to translate the information into a numerical code suitable for storage in a computer bank.

Methods and Material

A proforma was designed which recorded in numerical form the presence or absence of the many types of brain damage which can be sustained as a result of a head injury, *e.g.,* cerebral contusions, intracranial haematomas, hypoxic damage and diffuse axonal injury. Similarly, information was coded about extracranial injuries and the relevant clinical features. Where possible the lesions were quantified, *e.g.*, by stating the volume of intracranial haematomas, or the severity of cerebral contusions (assessed by the Contusion Index method[3]).

A detailed neuropathological examination was undertaken on 177 patients with fatal, non-missile head injuries and in each case the pathological findings were transferred to the proforma and then to a data bank. This initial step is time-consuming, but once completed it is then possible to obtain very rapidly and easily information about the frequency of any of the recorded features. Particular subgroups of the series can also be selected and the frequency of any lesion in that group compared with the remainder of the series.

Applications

This method has already proved useful in the field of head injury. It has been used to assess the validity of the contre coup hypothesis for cerebral contusions. The severity of contusions on each side of the brain was determined in a group of patients with localizing head injuries, and related to the site of the blow. The results obtained from the data bank clearly showed that contusions were *not* more severe contralateral to the site of injury[3]. When subjected to quantitative testing by the system described, it would therefore appear that the importance of the contre coup phenomenon has in the past been over-emphasized. The data bank has also helped to define more precisely the entity of diffuse axonal injury (DAI)[2]—a form of brain damage characterized by widespread damage to axons in the white matter of the cerebral hemispheres and brain stem. Our early experience had suggested that patients with DAI form a distinct clinico-pathological group. The data retrieval system was used to test this formally by comparing the clinical and pathological features of patients with DAI with the remainder of the series without DAI. The results showed very striking differences between the two groups[2], thus confirming that DAI is a distinct entity. This type of analysis is made possible by the ease of data retrieval and manipulation which is inherent in the computerized storage system.

Discussion

The system described therefore offers several advantages. In particular, it allows easy handling of information from large numbers of cases, thereby affording greater ease of data extraction and the presentation of results from a large series. It also allows more objective comparisons between groups by encouraging greater uniformity of description within a series. However, the limitations of the system must also be recognized. Access to computer facilities is a prerequisite and competent statistical advice is necessary to assist in data processing. Appropriate proforma design is also crucial because information can only be extracted from the bank if it has first been fed in. The proforma design must therefore anticipate the type of question which is likely to be asked of the system. Nor can the factor of human error be entirely eliminated since transcribing the information to the proforma involves interpretation of observed features and is thus a potential source of error. Nevertheless, provided the limitations are recognised, this system has a significant contribution to make to the storage and analysis of neuropathological data. It has already proved useful in the field of head injury[1], and it is hoped that this report will stimulate other workers to explore its use elsewhere.

References

1. Adams, J. H., Graham, D. I., Scott, G., Parker, L. S., Doyle, D., Brain damage in fatal non-missile head injury. J. clin. Path. *33* (1980), 1132—1145.
2. Adams, J. H., Graham, D. I., Murray, L. S., Scott, G., Diffuse axonal injury due to nonmissile head injury in humans: an analysis of 45 cases. Ann. Neurol. *12* (1982), 557—563.
3. Adams, J. H., Scott, G., Parker, L. S., Graham, D. I., Doyle, D., The contusion index: a quantitative approach to cerebral contusions in head injury. Neuropath. appl. Neurobiol. *6* (1980), 319—324.

Authors' addresses: Dr. Grace Scott, Medical Research Council Fellow; Miss Audrey Kerr, Research Assistant, University Department of Neuropathology; Mrs. Lilian S. Murray, Research Assistant, University Department of Neurosurgery, Institute of Neurological Sciences, Southern General Hospital, Glasgow, G51 4TF, Scotland, U.K.

Acta Neurochirurgica, Suppl. 32, 99—104 (1983)
© by Springer-Verlag 1983

*Queens University, Belfast; **Chemical Defence Establishment, Salisbury;
***Royal Army Medical College, London; and ****Maida Vale Hospital,
London, U.K.

Experimental Penetrating Head Injury:
Some Aspects of Light Microscopical
and Ultrastructural Abnormalities

By

I. V. Allen*, J. Kirk*, R. L. Maynard**, G. K. Cooper**,
R. Scott***, and A. Crockard****

With 2 Figures

Summary

A high velocity model of penetrating head injury has been developed in the
rhesus monkey and a lower velocity model in the baboon. It is apparent that
pathological changes are widespread and develop early although the pathogenesis
of the diffuse vascular changes is unknown. The present study involved the
sampling of grey and white matter from 20 monkeys with high velocity injury, and
10 baboons with low velocity injury together with similar material from a number
of normal control animals. 30 minutes after a high velocity injury swelling of
perivascular astrocytes was present, sometimes associated with an increase in
extracellular fluid. Animals with lower velocity injuries survived for some hours.
Astrocytic swelling and perivascular oedema associated with cellular necrosis was
frequently found in this group. The pathogenesis of these lesions is discussed.

Keywords: Head injury; astrocytes; oedema; electron microscopy.

Introduction

Penetrating head injury is a major cause of death and disability
in soldiers. Although methods of head protection from low velocity
bullets have been devised, no practical form of head protection

7*

from medium and high velocity bullets and fragments is yet available. Unfortunately injuries caused by these medium to high velocity projectiles are more severe and have a worse prognosis than those caused by low velocity ones. The pathology and pathophysiology of such injuries are poorly understood and have received scant attention in the scientific literature; they cannot simply be deduced by extrapolation from findings in non-penetrating or low velocity penetrating head injury. An understanding of these pathobiological processes may be of great value, both in the design of protective headgear and in the development of effective post-injury patient management and therapy. The few previous studies of head injuries from firearms in humans[6] and in experimental animals[5] are, in this respect, limited in value. As a further step in the investigation of such processes we have developed experimental models of both high velocity[3] and medium velocity[1] penetrating head injury in higher primates. The velocities sizes and masses of the projectiles were chosen so as to simulate closely the effects of modern infantry weapons on humans. The present brief report highlights the astrocytic response, an aspect of the pathological findings which may be of particular significance for pathophysiology and patient care. More detailed reports of the methodology and findings are being published elsewhere[1, 2, 3, 9].

Methods and Material

Details of the precise experimental and monitoring procedures are given elsewhere[1, 3, 9]. Briefly, in each series of experiments, the deeply anaesthetized animal was subjected to a transfrontal penetrating wound from a stainless steel ball. In the high velocity experiment using rhesus monkeys, a ball of 3.2 mm diameter impacted with the intact skull at a velocity of approximately 1,000 m/s. Medium velocity injury, to baboons, was caused by a 4.8 mm ball impacting directly onto the exposed dura at speeds ranging from 181 to 364 m/s. Fixation of the brains destined for pathological studies was by vascular perfusion with buffered formaldehyde or, more commonly, by buffered mixed formaldehyde and glutaraldehyde. Most animals were sacrificed within the first hour following the injury. A few baboons survived up to 4 hours before being sacrificed. White and grey matter was sampled from 20 monkeys with high velocity injury and from 10

Fig. 1. Early perivascular swelling following high velocity missile head injury. Rhesus monkey, toluidine blue. × 45

Fig. 2. Early swelling of pericapillary astrocytic end feet following medium velocity head injury. The endothelial cell and adjacent oligodendrocyte appear normal. Baboon, electron micrograph × 8,000

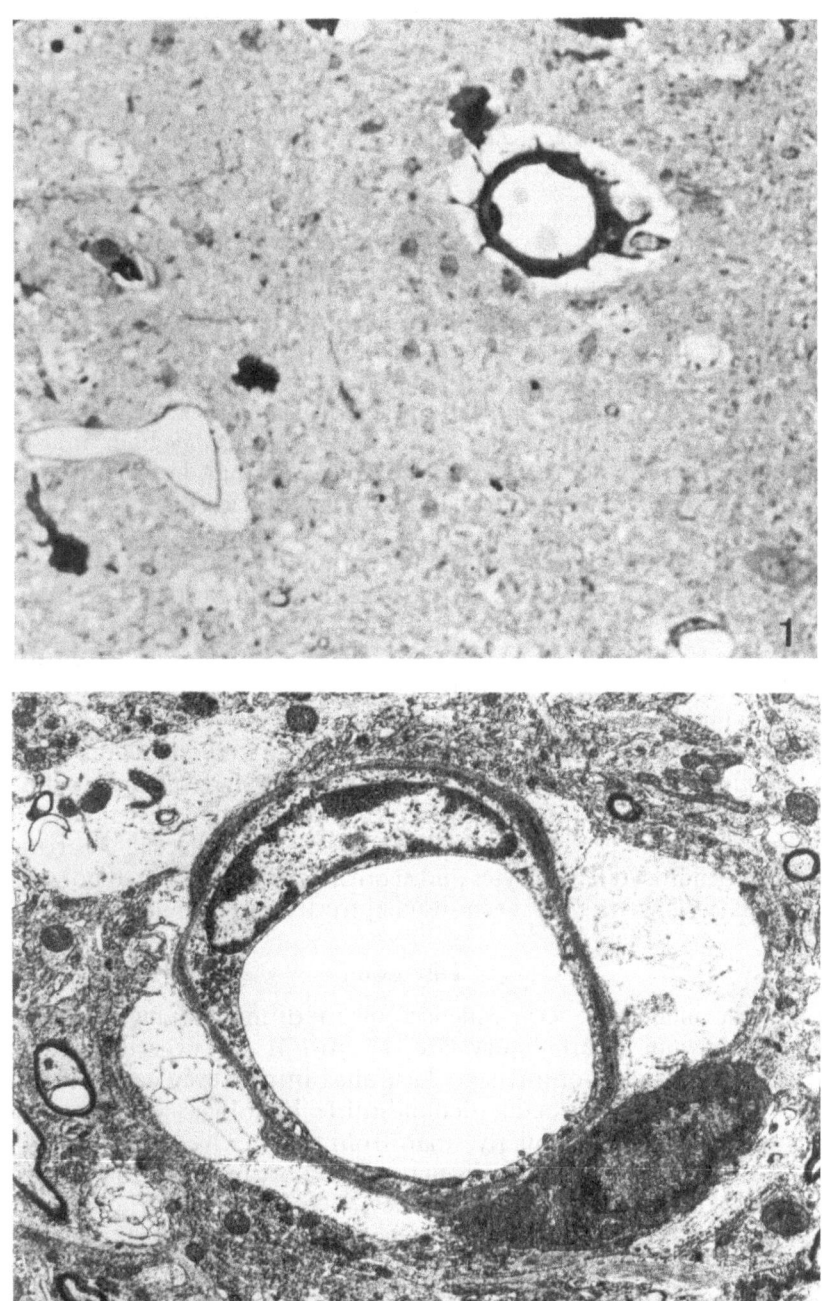

Figs. 1 and 2

baboons with medium velocity injury together with similar material from a small number of normal control animals of both species.

Specimen blocks were embedded in paraffin, low viscosity nitrocellulose, JB4 resin or Taab epoxy resin. Sections, stained by a variety of standard methods were examined by light and electron microscopy.

Results

All experimental animals showed macroscopical and histological abnormalities. In and immediately surrounding the wound track there was haemorrhage, disruption of tissue and extravasation of fluid. In the high velocity experiments bone fragments were also present in and around the missile track. More widely disseminated lesions were a constant feature at both missile velocities. In particular, extensive subarachnoid, intraventricular and subependymal haemorrhage were prominent. Perivascular ring haemorrhages, oedema, and blood vessels with surrounding zones of decreased or increased staining intensity were also common, both close to and away from the missile track. Such changes were seen particularly in deep white matter, hypothalamus, midbrain, pons and cerebellum. In the high velocity series, neuronal injury was not readily apparent. In the medium velocity series where long-term (4 h) survivors were examined, neuronal necrosis was evident within the wound track. Astrocytic swelling and disruption was a striking feature both inside and remote from the wound area. This swelling (Fig. 1), more evident in grey than in white matter, seemed selectively to involve astrocytes (Fig. 2). Adjacent neurons, oligodendrocytes, pericytes and endothelial cells appeared normal. Such swelling was not seen in uninjured control animals.

Discussion

The immediate consequences of medium and high velocity experimental head injury are, as might be expected, tissue disruption and haemorrhage. Neuronal injury, even in the wound track may not be readily demonstrable histologically until some time has elapsed. Similarly, apart from those which are physically disrupted, blood vessel walls may appear normal in the immediate post-wound period. Oligodendrocytes too, appear histologically normal. The finding of an astrocytic abnormality, namely swelling, very soon after injury is a new finding in brain injury. Although more prominent around blood vessels it also occurred perineuronally and elsewhere. As far as we are aware it has not been seen in human head injury, although Griffiths *et al.*[7] have reported it in experimental impact injury of the spinal cord. Jenkins *et al.*[8] have described it in complete cerebral ischaemia.

If, as we believe, such swelling is non-artefactual, then a number of questions arise. What is the physical or physiological mechanism responsible for its occurrence? What effect does it have on neuronal physiology, function and survival? Is it reversible in such injury-induced situations and would this aid or impede recovery? Bourke *et al.*[4] have reported a physiologically mediated and pharmacologically reversible astrocytic swelling in cat cerebral cortex. Does the same mechanism apply in head injury? We are not yet in a position to answer all of these questions but they are amenable to experimental investigation. In particular the possible breakdown of some blood-brain barrier functions, either as a cause or a consequence of astrocytic swelling may be worth studying[10]. To this end we are extending our work to included barrier assessment in animals with experimental penetrating head injury.

References

1. Allen, I. V., Crockard, A., Maynard, R. L., Cooper, G. K., Pathological changes following experimental medium velocity penetrating head injury. In: Neural trauma (Jane, J., ed.). New York: Raven Press. 1983 (in press).
2. Allen, I. V., Kirk, J., Maynard, R. L., Scott, R., Crockard, A., An ultrastructural study of experimental high velocity penetrating head injury. Acta Neuropathol. (Berl.) *59* (1983), 277—282.
3. Allen, I. V., Scott, R., Tanner, J. A., Experimental high velocity missile head injury. Injury *14* (1982), 183—193.
4. Bourke, R. S., Waldman, J. B., Kimelberg, H. K., Barron, K. D., San-Filippo, B. D., Popp, A. J., Nelson, L. R., Adenosine-stimulated astroglial swelling in cat cerebral cortex in vivo with total inhibition by a non-diuretic acylaryl-oxyacid derivative. J. Neurosurg. *55* (1981), 364—370.
5. Clemedson, C.-J., Falconer, B., Frankenberg, L., Jonsson, A., Wennerstrand, J., Head injuries caused by small-calibre, high velocity bullets: An experimental study. Z. Rechtsmed. *73* (1973), 103—114.
6. Freytag, E., Autopsy findings in head injuries from firearms. Arch. Pathol. *76* (1978), 215—225.
7. Griffiths, I. R., McCulloch, M., Crawford, R. A., Ultrastructural appearances of the spinal microvasculature between 12 hours and 5 days after impact injury. Acta Neuropathol. (Berl.) *43* (1978), 205—211.
8. Jenkins, L. W., Povlishock, J. T., Becker, D. P., Miller, J. D., Sullivan, H. G., Complete cerebral ischemia: An ultrastructural study. Acta Neuropathol. (Berl.) *48* (1979), 113—125.
9. Maynard, R. L., Cooper, G. J., Evans, V. A., Kenward, C. E., Pearce, B. P., Stainer, M. C., Aldous, F. A. B., Burdett, K. J., Howard, W., Allen, I. V., Crockard, A., Pathological effects of experimental medium velocity penetrat-

ing head injury. Chemical Defence Establishment, Internal Report, Ministry of Defence, 1983.
10. Povlishock, J. T., Becker, D. P., Sullivan, H. G., Miller, J. D., Vascular permeability alterations to horseradish peroxidase in experimental brain injury. Brain Research *153* (1978), 223—239.

Author's address: Prof. I. V. Allen, Neuropathology Laboratory, Department of Pathology, Institute of Pathology, Grosvenor Road, Belfast, BT12 6BL, U.K.

Acta Neurochirurgica, Suppl. 32, 105—107 (1983)

Department of Neuropathology, Aarhus Kommunehospital, Aarhus, Denmark

Midline Rupture of the Mesencephalon

By

Marie Bojsen-Møller

With 2 Figures

Summary

A midline rupture of the mesencephalon was found in 3 young males surviving closed head injury for 3–5 days. Other brain damage was relatively mild, but there was brain oedema with signs of herniation. In only one case were there symptoms of a hypothalamic lesion. The author suggests that a rupture is initiated by the compression of the brain stem against the clivus, whereby the pedunculi are displaced away from each other, and that the rupture naturally continues along the midline vessels to end in the aqueduct.

Keywords: Mesencephalon; rupture; trauma.

The finding of 3 cases of midline rupture of the mesencephalon makes it likely that this lesion may be mistaken for secondary brainstem damage or believed to be artificial in cases of instant death.

Material and Methods

Three males aged 15, 21, and 26 years respectively sustained severe frontal head impacts. The youngest was awake for 2 hours but the other two were rendered unconscious at once. One had a minor skull fracture, the two others had none. In only one patient were signs of a hypothalamic injury present. All three were treated with barbiturates and two of them underwent decompression surgery. Survival ranged from 3–5 days.

The autopsies revealed thoracic lesions in two and no organ damage in one.

The brains were fixed for 3 weeks, and routine staining procedures were performed on paraffin sections.

M. Bojsen-Møller:

Results

Almost identical lesions in the form of midline ruptures running from the interpeduncular fossa to the aqueduct were found in all three patients (Fig. 1). Oedema and minor secondary pontine haemorrhages were also seen. Histological examination of the midline lesions revealed distinct reactive changes which were older

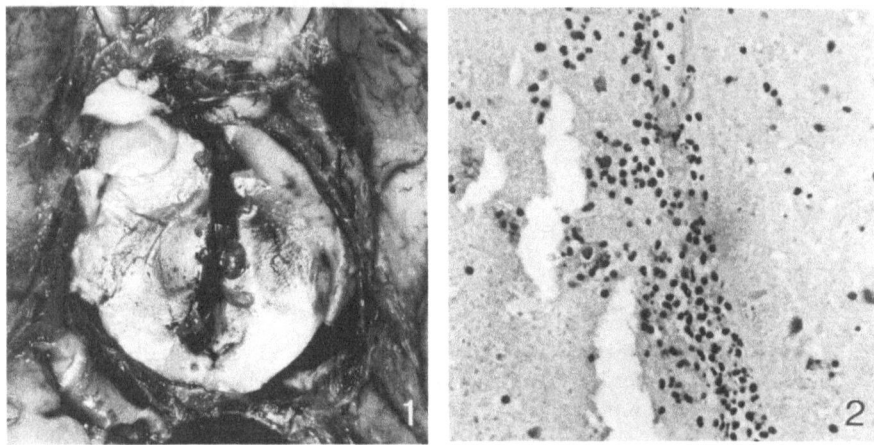

Fig. 1. Midline rupture of the mesencephalon. Male 26 years

Fig. 2. Vital reaction along a rupture. Male 21 years. × 100

than those seen around the secondary haemorrhages (Fig. 2). In an attempt to illustrate the mechanism of this type of damage an experiment was carried out on a corpse: the skull cap and the cerebral hemispheres were removed to expose the mesencephalon; a knife was then pressed against the dorsal surface of the brain stem resulting in deformation of the pedunculi which were easily flattened out against the clivus.

Discussion

The present three cases seem to conform to the characteristic description given by Crompton of patients with primary brain-stem lesions, being young males with closed head injuries and a short survival[2]. They are however different with regard to the localization of the lesions, and Crompton's cases seem to have hit the back of their heads whereas this paper deals with frontal impacts. He also mentions midline lesions as secondary brain-stem damage, but the present lesions are definitely primary.

To distinguish between primary or secondary brain-stem damage may be difficult[1], and it is easier to postulate that it is primary when no signs of increased intracranial pressure are found. The mere appearance of the ruptures and the obvious histological difference in the age of the lesions in the mesencephalon and in the pons in the present cases lead to the conclusion that both primary and secondary brain-stem lesions are present.

The author's conclusion is that a severe impact between the mesencephalon and the clivus may deform the pedunculi starting a rupture at the bottom of the interpeduncular fossa which then continues along the midline vessels to end in the aqueduct.

References

1. Adams, J. H., The neuropathology of head injuries. In: Handbook of clinical neurology, Vol. 23 (Vinken, P. J., Bruyn, G. W., eds.), pp. 35—65. Amsterdam: North-Holland. 1975.
2. Crompton, M. R., Brainstem lesions due to closed head injury. Lancet *i* (1971), 669—673.

Author's address: Dr. Marie Bojsen-Møller, Department of Neuropathology, Aarhus Kommunehospital, DK-8000 Aarhus C, Denmark.

Acta Neurochirurgica, Suppl. 32, 109—114 (1983)
© by Springer-Verlag 1983

Institute of Forensic Medicine of the Free University, Berlin

Brain-Stem Injury and Long Survival—a Forensic Analysis*

By

H. Bratzke

With 2 Figures

Summary

Out of 1781 autopsies undertaken on fatal head injuries over a period of 19 years (1960–1979) there were 387 cases with brain-stem lesions; 15 of these had survived more than one month.

Detailed macroscopic and microscopical studies showed different types of damage, which could not be explained by secondary processes only. The cases are demonstrated and discussed from the forensic viewpoint.

Keywords: Head injury; brain-stem injury.

Introduction

In the forensic assessment of blunt head injuries it is relevant to estimate the time, type and force of the trauma, as well as an individual's capacity to function after the accident. Brain-stem lesions play an important role in answering these questions.

Material and Methods

An evaluation of 11,591 forensic autopsies undertaken in the Institute of Forensic Medicine between 1960 and 1979[2] revealed brain-stem lesions in 387 out of 1781 head injuries (21.7%). 15 of these 387 cases (0.4%) had a survival time of more than 4 weeks.

---

* Supported by grants of the "Deutsche Forschungsgemeinschaft".

Case no.	Reg. no.	Age/Sex	Survival time	Type of trauma	Respiration	ICH*	Trepanation	Apallic Syndrome	Cachexia	Cause of death	Skull fracture	Hydrocephalus	Cortical lesion	Other fractures
1	526/79	60 y., ♀	43 d	traffic accident	(+)***	subd.	+	∅	∅	pneumonia	+	+	+	∅
2	604/72	44 y., ♂	46 d	fall (ground)	–	subd.	+	n.i.**	∅	pneumonia	+	+	+	∅
3	462/71	37 y., ♂	50 d	brawl	–	subd.	+	n.i.	∅	pneumonia	+	∅	+	∅
4	90/78	35 y., ♂	52 d	fall (high)	(+)	subd.	+	n.i.	∅	central dysregul.	+	+	+	+
5	576/80	15 y., ♀	54 d	traffic accident	–	∅	∅	n.i.	∅	pneumonia	+	+	+	+
6	201/79	59 y., ♀	60 d	traffic accident	–	∅	∅	∅	∅	pulm. emb.	∅	+	∅	+
7	150/76	34 y., ♂	66 d	traffic accident	–	subd.	+	n.i.	∅	pneumonia	+	∅	+	∅
8	537/71	7 y., ♂	85 d	traffic accident	(+)	subd.	+	n.i.	∅	sinus thrombosis	∅	–	–	+
9	718/78	51 y., ♂	90 d	traffic accident	(+)	subd.	+	+	+	pneumonia	+	+	+	+
10	63/81	25 y., ♂	111 d	traffic accident	–	subd.	+	+	+	pneumonia	+	+	+	+
11	18/74	31 y., ♂	117 d	traffic accident	–	∅	∅	n.i.	+	pneumonia	∅	+	∅	+
12	56/80	23 y., ♂	122 d	traffic accident	(+)	∅	∅	+	+	pneumonia	∅	+	∅	+
13	494/73	35 y., ♂	130 d	brawl	(+)	epid.	+	+	+	pneumonia	+	+	+	∅
14	223/77	20 y., ♂	133 d	traffic accident	n.i.	?*****	∅	n.i.	+	?	∅	?	?	+
15	203/69	44 y., ♂	168 d	traffic accident	–	∅	∅	+	+	pneumonia	+	+	∅	+

16	421/78	18 y., ♂	210 d	traffic accident	(+)	subd.	+	+	pneumonia	∅	+	+
17	525/79	4.5 y., ♂	214 d	traffic accident	(+)	∅	+	∅	pneumonia	∅	∅	+
18	182/80	25 y., ♀	1 y	traffic accident	(+)	∅	+	+	nephritis	∅	∅	+
19	232/79	49 y., ♂	1,5 y	fall (ground)	(+)	subd.	+	+	pulmonary embolism	+	+	∅
20	523/80	35 y., ♂	19 y	gunshot	(+)	∅	n.i.	+	malign. hyperth.	(+)	∅	∅

* ICH Intracranial haemorrhage.
** n.i. no information.
*** (+) spontaneous breathing recurring after artificial respiration.
**** Autopsy previously performed: corpse without internal organs.
∅ = absent.

Fig. 1. 20 cases with a survival longer than 1 month

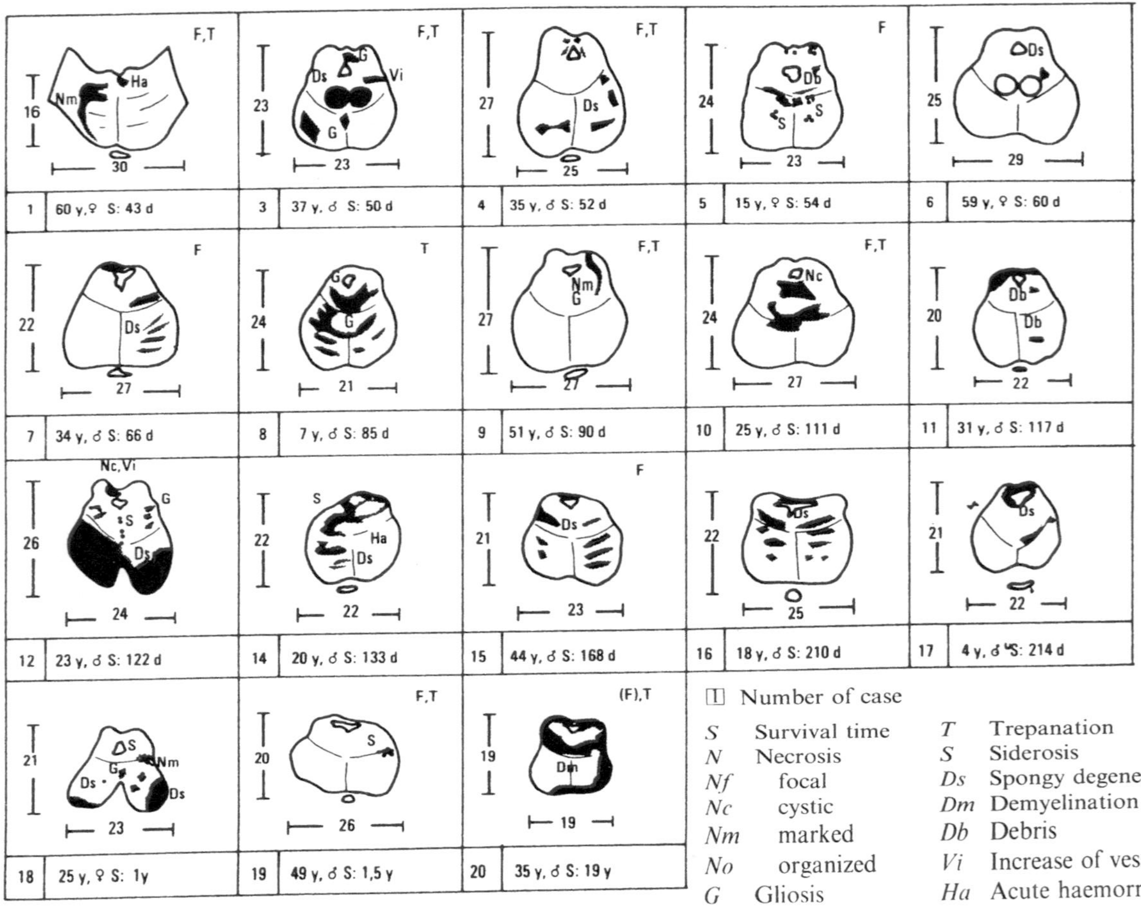

Fig. 2. Lesional pattern in the brain stem (diagrams from original slides, cases 2 and 13 are excluded)

□	Number of case
S	Survival time
N	Necrosis
Nf	focal
Nc	cystic
Nm	marked
No	organized
G	Gliosis
F	Skull fracture

T	Trepanation
S	Siderosis
Ds	Spongy degeneration
Dm	Demyelination
Db	Debris
Vi	Increase of vessels
Ha	Acute haemorrhage

Results

The main data in these cases as well as 5 other ones from 1980/81 are summarized in Fig. 1. Twenty cases are involved: 16 male, 4 female, and mostly young (15 were younger than 40). The survival time ranged between 43 days and $1\frac{1}{2}$ years, one gunshot wound having survived for 19 years (case 20). The distribution of the lesions in the brain stems are demonstrated (except for cases 2 and 13) by drawings of the original preparations (paraffin and celloidin) in Fig. 2.

Findings

A general feature was marked atrophy of the brain stem (reduction in width averaging 11%, and in depth 6%), and expansion and deformation of the aqueduct and 4th ventricle, sometimes with "aseptic ependymitis". Microscopical lesions consisted mainly of spongy long-tract degeneration in the pons and midbrain, particularly in the lateral and medial lemnisci, in the medullary velum and in the tectum of the mesencephalon. Occasionally there was only gliosis, and focal destruction of tissue. Siderosis was often conspicuous and in one case there were isolated pigmentary deposits (case 19). The formation of new vessels was seen only occasionally in necrotic areas, and was sometimes associated with acute haemorrhages (cases 1 and 14).

Conclusions

The 20 cases analyzed demonstrate the types of lesion found in some forms of protracted post-traumatic encephalopathy[3]. Whether the degeneration of the long tracts is a secondary phenomenon resulting from direct mechanical alteration[1, 5] or a form of "anoxic-vascular leucoencephalopathy", such as occurs with a lack of oxygen for various reasons[3] cannot be decided on the basis of this new material, especially since thorough knowledge of the clinical course was available only in individual cases. There was in particular no information on whether or not intracranial pressure had been high during life[1].

Our own experience suggests that rotational trauma has a particular tendency to produce anteroposterior extension and torsion of the brain stem and this may be of decisive prognostic importance even when additional cerebral injury is present[2, 4]. This is common in those cases in which there was no gross cranial injury or intracranial haemorrhage (cases 6, 11, 12, 17 and 18). Extensive damage is often attributable to secondary processes. However the cuneate necroses of the sulcus mesencephali, as in case 18, which cannot be viewed as "Kernohan's notch", support the concept of direct damage against the edge of the tentorium. Likewise, the sharply delineated foci of necrosis such as those seen in cases 1 and 9 are due to traumatic ruptures of vessels; the site of rupture however was not apparent although this is considered necessary for its proof[4].

The various types of lesion and reaction apparently depend on the duration of an anoxic phase and the extent of reperfusion. There is no evidence of correlation with the survival time.

Important clues are provided by the location of the deposits of haemosiderin which are the residue of earlier haemorrhages. Sometimes they correspond to secondary haemorrhages in the brain stem caused by congestion, in other cases to localized traumatic rupture of vessels.

Estimation of the time of trauma is complicated by the fact that it is not uncommon for fresh haemorrhages to occur in necrotic areas so that different stages exist next to each other (case 1). This might shed some light on cases of unexplained "sudden" death in a condition clinically considered "stable".

In forensic assessment, proof of a brain-stem lesion, whether of a primary or secondary nature, is often decisive, despite all of the difficulties involved, because that lesion may be the only irrefutable evidence of relevant brain injury. Obviously this is particularly important when the last physician to treat the patient erroneously certifies a "natural death" after long survival, although there was an injury at the beginning of the chain of events.

References

1. Adams, J. H., The neuropathology of head injuries. In: Handbook of clinical neurology (Vinken, P. J., Bruyn, G. W., eds.). Amsterdam-Oxford: North-Holland. 1975.
2. Bratzke, H.: Zur Kenntnis der Hirnstammverletzungen aus forensischer Sicht. Habilitationsschrift, Berlin 1981.
3. Jellinger, S., Seitelberger, F., Protracted posttraumatic encephalopathy. Pathology, pathogenesis and clinical implications. J. Neurol. Sci. *10* (1970), 51—94.
4. Krauland, W., Verletzungen der intrakraniellen Schlagadern. Berlin-Heidelberg-New York: Springer. 1982.
5. Strich, S. J., Diffuse degeneration of the cerebral white matter in severe dementia following head injury. J. Neurol. Neurosurg. Psychiat. *19* (1956), 163—185.

Author's address: Dr. H. Bratzke, Institute of Forensic Medicine, Free University, Hittorfstrasse 18, D-1000 Berlin 33.

Acta Neurochirurgica, Suppl. 32, 115—117 (1983)
© by Springer-Verlag 1983

Baylor College of Medicine, Houston, Texas, and Southwestern Institute of
Forensic Sciences, Dallas, Texas, U.S.A.

Head-In-Motion Contusions in Young Adults

By

J. B. Kirkpatrick

Summary

The pathogenesis of contusions was studied in a series of acute fatal closed head injuries. Important factors include: 1. whether the head was in motion or stationary; 2. the direction and magnitude of the force; 3. the presence of depressed fractures and lacerations; and 4. roughness of the overlying bone. In the young adult group, the frequent high velocity motor vehicle accidents create a dominant pattern of injury to the frontal and temporal lobes, usually sparing the occipital lobes and cerebellum.

Keywords: Head injury; contusions of the brain.

The patterns of tissue damage which occur in human brain trauma teach us much about the pathogenesis of injury. This information is lost, however, unless carefully organized observations are correlated with details of the accident. Such a series has been compiled from the medical examiner's office of Dallas County, Texas. The series comprises 55 males with acute fatal closed head injury, aged 16 to 30 years. The types of fatal accidents were: automobile driver—23, passenger—11, pedestrian—7, motorcyclist—5, plane crash—1, and falls—8. Significantly at this age, most of the accidents involved high speed motor vehicles, and falls were relatively less common than at other ages [3]. Alcohol was involved in 29 (53%) of these accidents.

The principal injuries were: scalp contusion—44, subgaleal haemorrhage—32, skull fracture—43, epidural haematoma—5, subdural haematoma—21, subarachnoid haemorrhage—37, contusion of brain parenchyma—55, parenchymal haemorrhage—16, cerebral oedema—23, secondary brain-stem haemorrhage—13, associated injuries of the body—23, and unrelated brain lesions—3. Lacerations were present in the cerebral cortex in 16 cases, in the corpus callosum in 5, in the superior cerebellar peduncle in 1, in the medulla or pons in 3, and in the pituitary stalk in 2. In order to use this series to study the pathogenesis of contusions, it is necessary, first, to exclude those cases which were complicated by laceration of the cerebral cortex, in addition to the contusion. These injuries invariably underly a skull fracture in which bone fragments have forcibly torn the brain. The damage results directly from the skull fracture, not merely from reverberations of the brain within the cranial vault[4].

Table 1. *Contusions of Brain Without Lacerations*

	Coup (direct) only		Both		Contre coup only
Frontal impact	8		6		0
Average severity (see text)		*3.4*		*3.2*	
Lateral impact	2		7		9
Average severity		*3.4*		*3.9*	
Posterior impact	0		1		5
Average severity		*3.0*		*4.0*	

The analysis of uncomplicated contusions is presented in Table 1. The direction of impact was determined by correlating information from the scene of the accident with lesions in the skin, subcutaneous tissue, and skull. Contusion of the brain was recorded as being present at the site of impact (coup lesion) or opposite (contre coup) or both. The relative size of contusions was recorded as a semiquantitative severity index: 2 + for lesions less than 1 cm, 3 + if between 1 and 3 cm diameter, 4 + if greater than 3 cm. With frontal impacts, contusions at the site of impact (*i.e.,* frontal and temporal poles) were more frequent, and more severe, than lesions at the opposite site. With lateral impacts, lesions at the opposite site (temporal or frontal lobe contralaterally) were moderately more frequent and more severe. Indeed, contusions at the impact site alone occurred in only two cases, while those at the opposite, contre coup site were frequent. Posterior impacts usually failed to injure the cerebellum or occiput, but always caused contre coup damage to the frontal and temporal lobes.

If contusions were caused entirely by a negative pressure phenomenon[5] this distribution would not be predicted. Instead, local constraints[2] and roughness of overlying bone[1] must play a significant role. The gyri have been shown to move in experimental animals, when the skull is struck[6]. These movements over the rough surfaces of the orbital plates and middle fossae create most of the injuries at these sites. The occiput and cerebellum are relatively protected by the smoothness of the overlying bone. The predominance of contusions at the coup site in frontal impacts to the moving head is also related to this phenomenon, and to fractures due to the extreme forces encountered in injuries in this age group.

References

1. Dawson, S. L., Hirsch, C. S., Lucas, F., Sebek, B. A., The contrecoup phenomenon. Reappraisal of a classic problem. Human Pathol. *11* (1980), 155—166.
2. Holbourn, A. H. S., Mechanisms of head injuries. Lancet *2* (1943), 438—441.
3. Kirkpatrick, J. B., Pearson, J., Fatal closed head injury in the elderly. J. Amer. Geriatrics Soc. *16* (1978), 489—497.
4. Lindenberg, R., Freytag, E., The mechanism of cerebral contusions. Arch. Pathol. (Chic.) *69* (1960), 440—469.
5. Lindgren, S. O., Experimental studies of mechanical effects in head injury. Acta Chirurgia Scand. Suppl. *360* (1966), 5—100.
6. Pudenz, R. H., Shelden, C. H., The lucite calvarium—a method for direct observation of the brain, II. Cranial trauma and brain movement. J. Neurosurg. *3* (1946), 487—505.

Author's address: Dr. J. B. Kirkpatrick, Professor of Pathology, Baylor College of Medicine, Department of Pathology, 1200 Moursund Avenue, Houston, TX 77030, U.S.A.

Acta Neurochirurgica, Suppl. 32, 119—123 (1983)

Neurological Institute, University of Vienna, Austria

Axonal Injury in Head Injury

By

P. Pilz

Summary

A histological analysis of 324 unselected fatal head injuries disclosed axonal injury in the form of retraction balls in 100 cases: this was severe in 64 and mild in 36. It is suggested that axonal injury exists as a spectrum without there necessarily being selective involvement of the corpus callosum or the rostral brain stem, and that cases with mild axonal injury may be unconscious for only a short time after their injury.

Keywords: Head injury; traumatic coma; axonal injury.

Introduction

The concept of diffuse axonal injury (DAI) as a primary traumatic lesion in the central nervous system was originally proposed by Strich[4] and expanded by Adams et al.[1]. It has, however, not gained general acceptance particularly in the German literature although these findings seem to constitute the changes repeatedly postulated by Spatz and his school as "trackless lesions of the brain stem". A major advance in this field has occurred only recently since it has been established that similar clinical and structural changes can be produced experimentally in subhuman primates using non-impact controlled angular acceleration of the head[2]. The present study was undertaken to establish the extent of axonal injury (AI) in head injury.

Methods and Material

Three hundred and twenty-four unselected cases of head injury were available for histological study. The brains were processed in the conventional fashion: the regions investigated histologically varied from case to case but in nearly every case at least two blocks from the brain stem were available. The sections were assessed particularly for the presence of AI in the form of retraction balls (RB) not associated with focal lesions. When at least a few unequivocal RB's were found, the case was classified as mild AI. When they were numerous and widespread, AI was classified as being severe.

Table 1. *Survival Time*

	Total	< 12 hours	12–24 hours	24–48 hours	2–7 days	1–4 weeks	> 4 weeks
AI present, n	100	0	5	8	39	37	11
AI absent, n	194	107	24	10	24	16	13

Table 2. *The Significant Differences Between Cases With and Without AI*

	AI present (n = 100)	AI absent (n = 194)	
Traffic accident	85	110	$p < 0.01$
Falls	9	54	$p < 0.01$
Lesion of anterior/superior cerebellar peduncle	38	19	$p < 0.01$
Central trauma (corpus callosum and basal ganglia)	34	31	$p < 0.01$
Gliding contusions	25	25	$p < 0.05$
Intracranial haematoma	20	65	$p < 0.05$

Results

Evidence of AI was found in 100 of the 324 cases: it was severe in 64 and mild in 36. The corpus callosum was affected in 79% of the cases with AI, the cerebral white matter in 52%, the internal capsule in 89%, the midbrain in 76%, the pons in 92%, and the medulla in 15%. In 48 cases both the cerebrum (the corpus callosum and/or the cerebral white matter) and the brain stem (the internal capsule and/or the midbrain, pons and medulla) were affected. In 10 cases AI was confined to the brain stem but AI was never restricted to the cerebral hemispheres alone. In 38 cases where data from the cerebral hemispheres was incomplete or not available, the brain stem was affected. AI was found in any age group but was rare under the age of 10 years (2 cases). The survival time of the patients

Table 3. *Patients with AI Are Grouped According to Concomitant Lesions in the Corpus Callosum (CC) and the Anterior/Superior Cerebellar Peduncle (a. c. ped.) to Make Them Comparable to the Series of Adams et al.*[1]

Group	n	Duration of unconsciousness							lucid interval	AI	
		< 5 minutes	5–60 minutes	1–24 hours	1–7 days	1–4 weeks	perma-nent	unknown		severe	mild
1 Lesion of CC and a. c. ped.	17	1*	—	—	—	—	16	—	1	17	0
2 Lesion of CC only	9	—	—	1	—	—	8	—	1	9	0
3 Lesion of a. c. ped. only	19	2	2	3	1	2	9	1	5	14	5
4 No lesion of CC or a. c. ped.	43	5	3	4	1	1	25	4	6	18	25

* Woman of 92 years, atrophic brain 970 g, only microscopic haemorrhages in corpus callosum.

is shown in Table 1, the differences between the cases with and without AI in Table 2, while in Table 3 the distribution of lesions in the corpus callosum and in the superior cerebellar peduncle (the dorsolateral quadrant of the rostral brain stem[1]) is presented.

Discussion

In this investigation AI was a frequent finding in unselected head injuries, occurring in 100 of 324 cases, and in nearly half of the patients who survived for more than 12 hours (100 of 212). RBs were found only in cases who survived for longer than 12 hours. The distribution of AI could not be assessed in every case since this was a retrospective study and only limited material was available in a proportion of the cases. Nevertheless AI in general seemed to be widespread. In some cases, however, it was not since in 10 only the brain stem was affected: in two of these 10 cases there was a subtotal ponto-medullary rent[3].

The differences in this series to that described by Adams et al.[1] are shown in Table 3. Thus in the 17 cases in group 1 there were lesions in the corpus callosum and in the superior cerebellar peduncle and AI was severe in them all. In contrast to the series of cases of diffuse axonal injury reported by Adams et al.[1] in which no case experienced a lucid interval, one of these 17 cases was unconscious for less than five minutes after injury. The reason for this is not known. In the cases in group 2 there was a focal lesion only in the corpus callosum but there was severe AI in every case. Groups 3 and 4 were clearly different in that there was a higher incidence of lucid interval and a short period of post-traumatic coma. There was also a higher incidence of mild AI. This would suggest that these patients sustained less severe brain damage and are similar to the less severe grades of DAI produced in subhuman primates by non-impact angular acceleration of the head[2]. Seven of the cases in groups 3 and 4 were unconscious for less than 5 minutes after their injury, and a further five for less than one hour (Table 3). It has, however, to be conceded that in a few cases RBs near focal infarcts brought about, for example, by tentorial herniation may have been mistaken for direct AI.

In conclusion it would appear that the criteria of DAI in man defined by Adams et al.[1] are indicative of its most severe form. This syndrome, however, occurs in only a small proportion of cases with some evidence of AI: thus in a large proportion of cases with AI there was no focal accentuation of the damage in the corpus callosum or in the superior cerebellar peduncle, while about one third of the cases of AI reported in this paper were not permanently unconscious after their injury.

References

1. Adams, J. H., Graham, D. I., Murray, L. S., Scott, G., Diffuse axonal injury due to non-missile head injury in humans: an analysis of 45 cases. Ann. Neurol. *12* (1982), 557—563.
2. Gennarelli, T. A., Thibault, L. E., Adams, J. H., Graham, D. I., Thompson, C. J., Marcincin, R. P., Diffuse axonal injury and traumatic coma in the primate. Ann. Neurol. *12* (1982), 564—574.
3. Pilz, P., Strohecker, J., Grobovschek, M., Survival after traumatic pontomedullary tear. J. Neurol. Neurosurg. Psychiat. *45* (1982), 422—427.
4. Strich, S. J., Diffuse degeneration of the cerebral white matter in severe dementia following head injury. J. Neurol. Neurosurg. Psychiat. *19* (1956), 163—185.

Author's address: Dr. P. Pilz, Landes-Nervenklinik Salzburg, Ignaz-Harrer-Strasse 79, A-5020 Salzburg, Austria.

Acta Neurochirurgica, Suppl. 32, 125—127 (1983)
© by Springer-Verlag 1983

Auckland Hospital, Auckland, New Zealand

Head Injury Unmasking Other Brain Diseases

By

W. E. Wallis and J. Wilson

Summary

Sixteen patients previously free of neurological complaints sustained minor head injuries and subsequently presented acutely with a wide range of unexpected, serious brain diseases. These included brain tumours, berry aneurysms, arteriovenous malformations and brain abscess. The possible, responsible mechanisms include direct mechanical trauma to the asymptomatic brain lesion, hydrocephalus and brain oedema.

This unusual complication of head injury should be suspected when a florid neurological syndrome follows a minor head injury. CAT scanning usually identifies the responsible disease. Because of its therapeutic implications, this complication deserves recognition in the differential diagnosis and management of head injuries.

Keyword: Minor head injuries.

Introduction

Scattered reports describe head injuries unmasking underlying neurological diseases which were previously asymptomatic[1-5]. Nevertheless, this phenomenon has not received systematic evaluation and is only rarely mentioned in the head injury literature. This paper describes 16 patients with this phenomenon and two further patients with known brain disease who had CAT scanning before and after a subsequent head injury.

Methods and Material

The 16 patients who were asymptomatic prior to the unmasking of their brain disease by head injury fulfilled five criteria.

1. There were no symptoms referable to underlying brain disease prior to head injury.
2. The head injury brought to medical attention and was not caused by underlying brain disease.
3. Neurological symptoms developed immediately or within one week after the head injury.
4. The neurological symptoms after head injury were compatible with the subsequently verified brain disease.
5. The nature of the underlying brain disease was verified by CAT scanning, angiography and neurosurgery.

Two further patients, with known brain disease, deteriorated after minor head injuries. These patients had CAT scanning before and after the head injuries.

Results

In the group of 16 previously asymptomatic patients the head injuries were generally minor and produced severe and unexplained neurological symptoms not compatible with the head injury alone. They were subsequently found to have a wide range of unexpected, serious brain diseases including seven brain tumours, four berry aneurysms, four arteriovenous malformations and one brain abscess. The two other patients who had known brain disease prior to their head injury had a cerebral oligodendroglioma and a giant posterior fossa aneurysm. Both patients had CAT scanning and did not have enlarged ventricles. Both deteriorated markedly after minor head injuries and were subsequently shown to have hydrocephalus on repeat CAT scanning. Both improved with shunting procedures.

Discussion

The failure to recognize minor head injuries unmasking previously silent neurological diseases may lead to inappropriate management of such patients. This condition should be suspected if patients develop persistent and dramatic neurological complaints and findings following minor head injuries. CAT scanning is generally the diagnostic procedure of choice. A single mechanism is unlikely to be responsible for this phenomenon, although post-traumatic hydrocephalus was shown to be present in at least two patients. Nevertheless, hydrocephalus was not present in all patients and other mechanisms might include direct mechanical trauma to the asymptomatic brain lesion and brain oedema.

References

1. Blau, A., Richardson, J. C., Strokes and head injury. Can. J. Neurol. Sci. *5* (1978), 263—266.
2. Levine, D. N., Black, P. McL., Kleinman, G. M., Acute quadriplegia due to cerebral metastases. Neurology (N.Y.) *31* (1981), 343—346.
3. Sadik, A. R., Adachi, M., Ransohoff, J., Rupture of an intracranial aneurysm within the subdural space—in association with trauma. A case report. J. Neurosurg. *20* (1963), 609—612.
4. Shealy, C. N., Craniopharyngioma diagnosed after head trauma. Arch. Neurol. (Chic.) *13* (1965), 217—218.
5. Varma, T. R., Sedzimir, C. B., Miles, J. B., Post-traumatic complications of arachnoid cysts and temporal lobe agenesis. J. Neurol. Neurosurg. Psychiat. *44* (1981), 29—34.

Authors' addresses: Dr. W. E. Wallis, Department of Neurology; Dr. J. Wilson, Department of Radiology, Auckland Hospital, Auckland, New Zealand.